*Minerals
and Man*

Photographs by

Studio Hartmann
Emil Javorsky
Katherine H. Jensen
Reo N. Pickens Jr.
John H. Gerard
Benjamin M. Shaub
Tad Nichols
and others

A Chanticleer Press Edition

Minerals and Man

by Cornelius S. Hurlbut, Jr.
Professor of Mineralogy, Harvard University

RANDOM HOUSE, NEW YORK

The publishers and author thank E. P. Dutton
& Co. Inc. for permission to quote from "On the
Vanity of Earthly Greatness" from Gaily
the Troubadour by Arthur Guiterman, and Houghton
Mifflin Company for permission to quote from
"The Dorchester Giant" by Oliver Wendell Holmes.

Third Printing 1970

Published in New York by Random House, Inc.
and distributed in Canada by Random House of
Canada Limited, Toronto

Planned and produced by Chanticleer Press, New York

Manufactured in Zurich, Switzerland:
Color separations by Cliché + Litho
Color printing by Conzett & Huber
Black and white gravure printing by Conzett & Huber

Library of Congress Catalog Card Number 68-28329

Contents

Preface 7
1 Minerals and How to Know Them 8
2 Early Use of Minerals 26
3 The Earth and Its Rocks 37
4 Minor Minerals of the Rocks 60
5 Nature's Treasure House 71
6 A Traprock Suite: Zeolites and Related Minerals 92
7 Crystals in Sedimentary Rocks 103
8 Minerals from the Sea 118
9 Minerals of Land-Locked Lakes 131
10 Metals as Minerals 142
11 Ore Minerals of the Common Metals 159
12 Minerals That Glow in the Dark 181
13 Iron and Its Ores 187
14 Minerals of the Precious Stones 208
15 Quartz 227
16 Colored Stones and Ornamental Minerals 249
17 Minerals for Atomic Energy 264
18 The Mines and Minerals of Cornwall 272
19 An Incombustible Fabric and a Stone That Burns 283
Appendix 293
Index 297

Preface

A fine collection of minerals often causes one to marvel that objects of such beauty with their subtle variations in color and symmetry of crystal form are products of nature. It was these same characteristics that attracted early man to minerals thousands of years ago and caused him to use them for his personal adornment and for their presumed magical powers. But minerals had a far greater importance to man than their aesthetic appeal, and throughout his long evolutionary period he has sought and used them for his weapons and personal comforts. With each succeeding generation man's dependence on minerals has become greater. They make our modern civilization possible, for all inorganic materials of commerce that are not minerals themselves are mineral in origin.

Early in the twentieth century it became possible to determine the internal structure of crystals by means of x-rays, and since that time the science of mineralogy has undergone revolutionary changes. Many of the natural science aspects have given way to the more technical and restricted investigation of mineral properties in the light of their internal atomic arrangement. There is thus a tendency in the modern approach to neglect a fascinating chapter in earth science, the origin, occurrence, association and uses of minerals. It was these aspects of mineralogy that formed the most memorable part of my first college course. And since that time I have cherished the hope that some day I would write a nontechnical book recounting the stories of minerals that enthralled me as a student, and adding to them some of the exciting accounts of mineral discovery and exploitation.

This book is thus the fulfillment of a longstanding ambition, and with its many and superb illustrations far exceeds the book I originally envisaged. For making it possible I am greatly indebted to Chanticleer Press and the members of its staff. Especial thanks are due Mr. Milton Rugoff for his skillful pen in sharpening ideas and eliminating superfluous words; to Mrs. Constance Sullivan for her patient world-wide search for appropriate photographs; and to Mr. Ulrich Ruchti for the attractive layout and design.

Originally this book was to have been written under the joint authorship of Dr. Henry E. Wenden and myself and it was to my deep regret that Dr. Wenden was unable to see the work through to completion. Nevertheless he must be credited with much of the material throughout and particularly that of Chapters 2, 5, 6 and 8.

Finally I wish to express my thanks to Dr. Arthur Montgomery who read and criticized the manuscript with the eye of a professional mineralogist; and to my wife, Margaret R. Hurlbut, who read it to determine whether it would be intelligible to the layman.

CORNELIUS S. HURLBUT, JR.
Cambridge, Massachusetts

1 Minerals and How to Know Them

In all probability our planet was at one time a fiery ball that gradually cooled as heat escaped into outer space. When a hard surface eventually formed, minerals made up the primordial crust. For a period of more than four billion years since then, a continuous succession of minerals has formed only to be broken down mechanically or destroyed through melting or chemical reactions. The elements which composed them persisted, formed new minerals and began another cycle. Thus minerals constitute the most substantial link with the earth's past, and only through them can we interpret the long and complex history of the earth.

The rocks that form the earth's crust are composed of aggregates of minerals mostly in small nondescript grains. But occasionally larger grains in the form of sharply faced crystals of various shapes and colors turn up in rocks. It is largely these rarer and more unusual minerals, many of them striking in appearance, that are dealt with in this book. Not only are they interesting to the mineralogist and mineral collector today, but they have fascinated man since he began to take note of his surroundings. From the very beginning man has depended on minerals for his weapons, his adornments and his comforts. Since the use of minerals is as old as civilization it is surprising that mineralogy as an integrated science did not develop until relatively recent times.

Mineralogy is not a fundamental science but rather a synthesis in which chemistry, physics and mathematics are used to explain minerals as a part of geologic history. It is thus a natural science without laws and generalizations of its own; and, because of the complexity of its materials, it is primarily descriptive. Nevertheless, mineralogists make observations as precise as those of the physicist, chemist or astronomer. With high precision they routinely measure

Irregularly intergrown cubes of purple fluorite from Illinois have crystallized on a surface of chert.
(Studio Hartmann)

the position of atoms in mineral crystals and the distance between them, as well as optical, mechanical and electrical properties. The environments in which minerals form are highly complex, far more so than those in laboratory experiments. However, experimental mineralogy is one of the chief modern-day resources of this science and by means of controlled experiments the mineralogist is able to describe, at least in part, the conditions under which minerals have formed in nature.

The central concept of mineralogy today is the principle that the external forms, physical properties and chemical behavior of crystals, are all the result of two fundamental factors: chemical composition and internal structure. On them depend the complex of optical, mechanical, electrical and chemical properties that characterize each mineral. This concept, in its bare form, is very old, and the story of its growth to fullest scientific meaning is essentially that of the science of mineralogy.

During the centuries of confusion and disorder that followed the collapse of Roman power in the West, the little mineralogical knowledge contained in such works as those of Pliny and Theophrastus was ignored or forgotten. It was thirteen hundred years before man again gave thought to the origin, classification and structure of minerals. But in every age there has been a subterranean mineralogy so to speak, a technical mastery of materials necessary for the conduct of the mineral industries. The possessors of this large body of mineral lore were simple, generally illiterate folk who wrote no books and framed no theories. Their living was obtained by working with minerals, and hence they preserved, practiced and passed on to their children or apprentices their hard-won arts.

This mineral technology was ignored by, or indeed was quite unknown to, the scholars who wrote books or made learned compilations. Further, mining was held to be a base occupation, degrading those who engaged in it. Thus under the double bar of ignorance and prejudice, mineralogy had little chance of recognition as a suitable subject for students. In spite of these handicaps, some works on minerals were published in every century.

Since the early writers had little practical contact with minerals, their works contained much lore and fable and only traces of original observation or critical examination of source materials. For example, in the thirteenth century, Albertus Magnus in his encyclopedic "Natural History" of which five books were devoted to minerals, treats as "valuable" those stones that bestow supernatural powers on the owner. He devotes space to explaining why one stone can make a man invulnerable and another free him from drunkenness and he implies that he knows from his own experience that "draconites" is a stone that comes "from the head of a dragon." Although very inaccurate and lacking in original information, Albertus' work was the best in its time and gives us a fair idea of mineralogical thought in the latter part of the Middle Ages.

If one were to select a single event as the end of the dark ages of mineralogy and its beginning as a science, it would be the publication of *De Re Metallica*. Written by a German physician, Georgius Agricola, in 1546 and translated in 1912 from the Latin by Herbert Hoover, former President of the United States, and his wife, this work gives an accurate description of the mining practices of the time and includes the first truly factual account of minerals. The next milestone in the science was based on observations of crystals of quartz made

*(Above) Amber-colored calcite from Indiana displays the crystal forms of scalenohedron and rhombohedron.
(Reo N. Pickens, Jr.)*

*(Below) Ore from a lead-zinc mine in Kansas shows a succession of minerals deposited on chert: small brilliant crystals of sphalerite followed by crystallization of well-formed cubic crystals of galena and finally by white to colorless crystals of calcite.
(Studio Hartmann)*

in 1669 by Nicolaus Steno. After examining crystals from many localities, Steno noted that despite their differences in size, shape and origin, the angles between corresponding faces were constant. By establishing that external form is not a matter of chance but a characteristic property of a mineral, he pointed the way for the study of other crystalline substances. Today the student learns as Steno's law the generalization that *the angles between equivalent faces of crystals of the same substances are constant.*

A century later in a book published in 1784, René Just Haüy, professor of mineralogy and crystallography in Paris, presented a revolutionary view of crystals. He saw them as built of minute identical units, "integral molecules," stacked in a regular manner. He further showed how the arrangement of these building blocks could account for the smooth faces of crystals. Although Haüy's "integral molecule" is no longer accepted, his ideas laid the groundwork for modern concepts of crystals and earned him the title of "Father of Crystallography." During the nineteenth century mineralogists concentrated on elaborating earlier ideas and major advances in the techniques of mineral study were developed. These advances included the optical goniometer, an instrument used for measuring crystal angles accurately; the Nicol prism, used in studying optical properties; and new methods for making chemical analyses of minerals.

Finally, in 1912, an experiment suggested by Max von Laue of the University of Munich opened the way to a new era in mineralogy. It was found that x-rays passing through a crystal were "reflected" from atomic planes and on emerging yielded information regarding the internal structure. Since then, x-rays have assumed an increasingly important role in the study of minerals; today their use is routine in every mineralogical investigation. With them one can determine the size and arrangement of the atoms within a crystal, the distance between atomic planes and the dimensions of the building unit, the unit cell. No other tool has given us so much information, for with a knowledge of mineral structures has come an understanding of the easily determined properties of minerals.

What Is a Mineral?

Minerals are products of nature that existed on the earth long before life made its appearance. The mineralogist thus studies only natural substances that are inorganic and excludes those that are the direct result of life processes. Consequently, coal and petroleum, two of the most important "mineral resources," are excluded because they are of organic origin. For the same reason the pearl is not included even though it has all the properties of the mineral aragonite. In addition to being natural and inorganic, a mineral must meet another requirement: it must be a chemical element or compound. It cannot be a random mixture of elements; the atoms that make it up must have definite ratios to each other, so that its composition can be expressed by a chemical formula. Not only are the proportions of the various atoms of a given mineral fixed, but so are their relative positions. These attributes give to each mineral a set of properties that characterize it so uniquely that one can distinguish it from all other minerals. The remainder of this chapter describes these properties and tells how they can be determined.

An illustration from Agricola's De Re Metallica *(1546) shows ore mined underground, moved through horizontal tunnels and hoisted up vertical shafts.*

Crystals of Minerals

Crystals with their flashing faces and the beauty of geometrical regularity have most fittingly been called the flowers of the mineral kingdom. Their smooth surfaces, resembling the facets of polished gems, often excite wonder. It is not uncommon to hear someone viewing a display of crystals of minerals exclaim, "they look as though they had been cut and polished." As far back as the first century A.D., Pliny the Elder thought that Indian gem cutters had been responsible for shaping beryl into hexagonal prisms. Yet the faces of crystals are a natural growth—the external expression of the orderly arrangement of the atoms within.

Even students familiar with the wonders of crystal growth gasp when they first see a two-hundred-pound ice-blue topaz from Brazil covered with glittering, mirror-like faces as large as the palm of a hand. Equally arresting are clear Brazilian quartz crystals weighing several hundred pounds, gypsum blades several feet in length from Naica, Mexico, and apatite crystals as big as tree trunks from the Halliburton-Bancroft district of Ontario. The giants in museums are far from the largest of their kind; others are too large to be removed from the rocks in which they are found.

In a few areas, large crystals can be seen in place, as in the Black Hills of South Dakota; embedded in the quartz and feldspar rock of the wall of the now unworked Etta mine are "logs" of spodumene, crystals of natural lithium silicate more than ten feet long and a foot across.

Feldspar undoubtedly forms the largest crystals, and today in many parts of the world one can see at least remnants of these giants several feet across on the walls of abandoned quarries. North of Kristiansand, Norway, a crystal was described with dimensions of $7 \times 12 \times 30$ feet; and in the Ural Mountains a quarry 30×30 feet of unknown depth was opened in a single feldspar crystal. Such large crystals are usually poorly developed because of interference from neighboring minerals during their formation, and smaller specimens thus often make better exhibits. It is safe to say that the smaller the crystal, the more perfect its development and the more faces it will possess. For this reason many collectors specialize in specimens of tiny crystals of minerals cemented into small boxes for display. The true beauty of these micromounts can be appreciated only when observed under a microscope.

Crystal Systems

Although we rarely encounter crystals with smooth shining faces in our daily round, we are constantly exposed to crystalline materials: the cup from which we drink our coffee, the sugar used to sweeten it, and the spoon with which we stir it are made up of atoms arranged in the orderly fashion of the crystalline state. If we were to x-ray these materials, the diffracted beams falling on photographic films would reveal dark spots in geometric patterns. The patterns would all be different from each other but each would be characteristic of the substance that produced it. From the positions and intensities of the spots we can determine the arrangement of the atoms and calculate the precise dimensions of the unit cell.

The unit cell, the smallest subdivision of a crystal, is a regular arrangement

Isometric

Tetragonal

Orthorhombic

Monoclinic

Triclinic

Hexagonal

of atoms held together by electrical forces. In a general way it is like the integral molecule described by Haüy, a building block that is too small to be seen, but when repeated over and over in space builds up a crystal as bricks build up a wall. The dimensions of the unit cell vary from mineral to mineral, depending on the kinds, number, size and arrangement of the atoms in it. An almost infinite number[1] of these building blocks must be stacked in a three-dimensional array to form a crystal, or even crystalline grains, large enough to be visible. Since there can be no space between the blocks, their shape is limited to only six basic forms.

According to the shape of the unit form, crystals are grouped into six crystal systems as follows:

> 1. *Isometric system.* This is sometimes called the cubic system since the units are cubes with the three edges equal in length and at right angles to each other.
> 2. *Tetragonal system.* The three edges of the units are at right angles to each other but only two edges are of equal length.
> 3. *Orthorhombic system.* The three edges are at right angles to each other but all are of different lengths.
> 4. *Monoclinic system.* The edges are all of different lengths; two of them are at right angles to each other but the third is inclined to the plane of the other two.
> 5. *Triclinic system.* The three edges are of different lengths and they are not at right angles to each other.
> 6. *Hexagonal system.* Two edges are equal and make angles of 60° and 120° with each other. The third edge is at right angles to them and of different length. Three of these units fitted together form a regular hexagonal prism.

Occasionally, well-formed crystals of minerals resemble the units of the six crystal systems but more commonly they are modified by the presence of other faces. These additional planes are not haphazard but, as observed by Steno long ago, have a constant angular relationship that is determined by the dimensions of the unit cell. A simple experiment to illustrate the principle of fixed angles can be carried out with children's building blocks or with dominos. Let us assume that we have 125 dominos, representing identical unit cells of the orthorhombic type. Stacking these together with corresponding edges parallel, we can build a structure with five units on each edge, that is, with the same proportions as a single unit but with all dimensions increased five times. If we take away one row of dominos from the upper right and upper left edges, we get two "crystal faces" truncating the edges of the fundamental form. If strips of cardboard are laid on the edges of the dominos on each side, the surfaces formed will make equal angles with the top surface of the pile. If we remove two front-to-back rows from both the right and left sides of the top layer, leaving only the center row, and then remove one row from both sides of the

---

[1] Ordinary salt, sodium chloride, has four atoms of chlorine and four atoms of sodium in each unit cell. In a grain of the size used for table salt there are about 5.6×10^{18} unit cells; the number, perhaps more familiarly expressed, is: 56 followed by seventeen zeros!

second layer, the two "crystal faces" will now be larger, consisting of flights of steps, but the angles between them and the top surface will remain as before.

It is easy to illustrate other possible "faces." For instance, if we remove the corner dominos from the top layer, four "faces" will be developed all of which make the same angle with the top surface. The faces on real crystals are formed by the omission of unit cells during growth rather than by the removal of such cells. Crystals which have corresponding faces of exactly the same size are sometimes described as "perfect." Such crystals have a greater geometrical symmetry than those which, through an accident of growth, have unequally developed corresponding faces, but they are basically no more perfect. Crystal perfection lies in the regularity of internal structure that is expressed not in the size of faces but in their angular relationship.

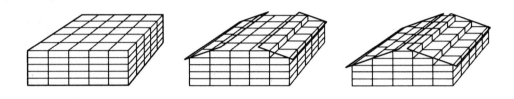

Stacking of dominos illustrates the stacking of unit cells in crystals. The removal of one row from both the upper right and upper left edges develops two "crystal faces" having fixed angular relations to the top surface. On the removal of additional rows of dominos, larger "faces" are formed but the angles remain the same.

One of the charming properties of crystals is their ordered beauty. This is evident not only on x-ray film but to anyone who holds a crystal of quartz, barite, fluorite or topaz in his hand, for in addition to being smooth and lustrous, the faces have a pleasing symmetry of arrangement. By means of this symmetry, reflecting the unit form, a crystal can be assigned to one of the crystal systems.

Crystal Symmetry

Crystals possess three basic kinds or "elements" of symmetry: symmetry planes, symmetry axes and symmetry centers. Some crystals have several planes and axes, others have very few or none; it is according to these that crystals are classified.

The easiest type of symmetry to understand—probably because human beings themselves have it—is symmetry with respect to a plane. Many objects, such as a square table, a cigar box or a chair, have this symmetry. An object has a symmetry plane if, were it cut in two, a mirror placed against the cut surface would produce an image that appears to restore the complete object. A symmetry plane, then, divides a crystal into two halves so that one is the mirror image of the other. Some crystals have no such symmetry plane but most have one to nine (except for eight which other symmetry elements do not permit).

The second symmetry element, axis, is an imaginary line about which an object is rotated. The number of times the object repeats itself in space during a complete rotation determines the kind of axis. A line through the center of a square table is a fourfold axis, for the table repeats itself four times when rotated about it; a six-sided pencil has a sixfold axis parallel to its length; and a match box has three twofold axes passing through the centers of opposite

(Right) Perfect octahedral cleavage fragments of fluorite illustrate the range of color in which the mineral is found. (Katherine H. Jensen)

16

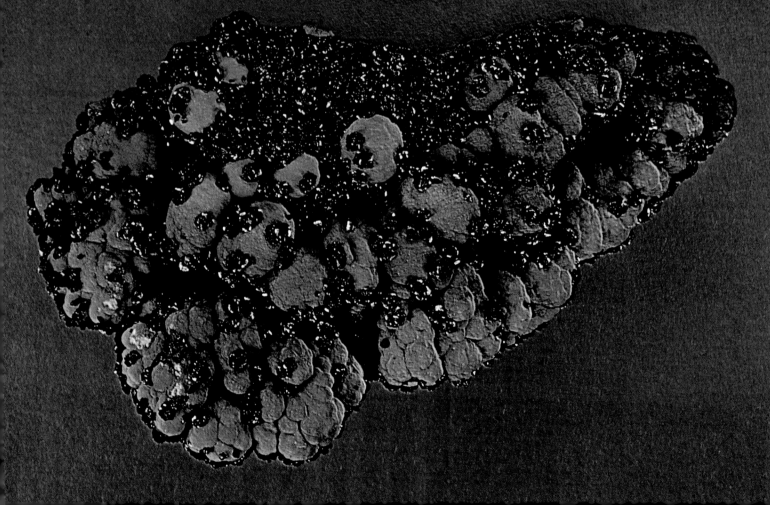

A symmetry plane is illustrated by placing half a crystal against the mirror. The image produced by reflection appears to restore the complete crystal.

A match box and a crystal both have three symmetry planes and three two-fold symmetry axes.

Although greatly different in appearance, a monument, a square table and a crystal have common symmetry elements. All have four planes of symmetry and a four-fold symmetry axis.

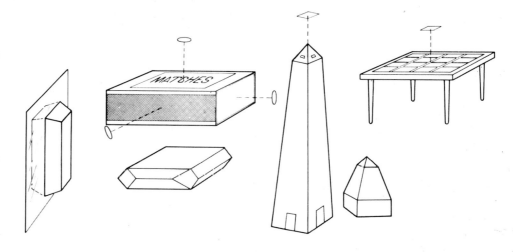

(Left above) Blue, tabular barite crystals on a base of red hematite from Cumberland, England. (Emil Javorsky)

(Below) Botryoidal malachite from Bisbee, Arizona, is encrusted with crystals of blue azurite, a common association. (Katherine H. Jensen)

sides. If the symmetry axis has a plane of symmetry at right angles to it, the two ends are similar and interchangeable; but if no such plane is present, as in the table or the pencil, the opposite ends are different and the axes are said to be polar. There are two, three, four and sixfold symmetry axes in crystals, but some crystals have none.

The third symmetry element is symmetry center. It is present if an imaginary line can be passed through the center of a crystal from any point on the surface to a similar point on the opposite side. The match box thus has a center of symmetry but the table does not. Parallel opposite faces on crystals usually indicate a symmetry center.

There are only thirty-two possible combinations of axes, planes and center of symmetry in crystals. These are known as the crystal classes, including one that has no symmetry whatever. The question naturally arises: If all crystals are built up by repetition of one of the six fundamental forms shown on Page 15, why are there more than six crystal classes? The answer lies in the arrangement of atoms within the unit cell. The different arrangements are reflected in the geometry of the faces and thus the external symmetry of the crystal.

Physical Properties of Minerals: Color

We can observe certain properties of a mineral merely by looking at it: the regularity of crystal form, the color and the luster. If asked to classify an assortment of minerals, a beginning student will generally arrange them according to color and luster.

The color of minerals is at once the most mutable and most constant, the most observed and perhaps least understood of mineral properties. For the ancients the mystical power of stones lay partly in their color, and the charm of gems still depends largely on color. Neanderthal men sprinkled their dead with powdered hematite in an effort to induce by sympathetic magic the ruddy color of life. Minerals have been used as pigments almost as long ago as they have been used as tools and weapons. Some minerals possess colors as constant and characteristic as their chemical compositions. Thus, malachite is always green, azurite and lapis lazuli are always blue, and rhodonite is always red or a red-brown. These colors are associated with a chemical element whose ions impart strong colors, whether in solution, in glass or in minerals.

Since most minerals are made up of elements which produce no characteristic color, they are colorless. Yet colored varieties of these minerals are known and are frequently the most common. In some cases the color results from appreciable amounts of elements, such as iron, that have strong pigmenting properties. In others the cause of the color is more subtle and the impurity elements are disproportionately small compared to the depth of color they produce. For example, in comparing two varieties of quartz, water-clear rock crystal and deep purple amethyst, chemical analyses reveal no iron in the rock crystal and only a trace of it in the amethyst. A possible explanation for the color difference is that iron atoms, not normally present in quartz, have damaged the crystal structure of the amethyst, permitting the absorption of certain wave lengths of light that usually would have passed through. Such minute impurities are responsible for the color of many gems, such as the blue of sapphire, the red of ruby and the green of emerald.

During growth, minerals may incorporate minute quantities of other minerals that color the host. Thus quartz may appear green through the presence of chlorite, and calcite may be made red by particles of hematite. One of the first lessons in the study of minerals is discriminating between minerals in which color is inherent and those in which it is vicarious.

Luster

Minerals can be sorted out easily and quickly on the basis of luster, which is sometimes defined as the quality of shining in reflected light. Those that reflect light as a metal does are said to have a metallic luster, whereas all others, including most common minerals, have a nonmetallic luster. In general, metallic minerals are opaque and since they reflect most of the light falling on them, they have a brilliant appearance. Nonmetallic minerals are most often light-colored, and since they permit light to enter them, less is reflected from the surface. Some are quite transparent and light passes through them as easily as through clear glass. In fact, many of them appear glassy on crystal faces or fracture surfaces and are therefore said to have a vitreous luster. However, the luster of some transparent minerals is hard and brilliant, like that of a diamond, and is therefore called adamantine.

When a beam of light enters a transparent substance it is refracted, that is, it is bent and follows a different path. The refractive power, different for each substance, is expressed by a number called the index of refraction. Since the refractive index of minerals can be measured with extreme precision, it is an important property in mineral identification. Refractive index measurements are one of the principal methods used by jewelers for gem determination; and they can be made while the stone is still in its mount.

Luster is related to refractive index, and both properties depend on the kind of atoms present and their structural arrangement. A high refractive index and an adamantine luster are characteristic of transparent minerals containing heavy metals. Examples of this among common minerals are cerussite (containing lead), cassiterite (tin), scheelite (tungsten) and cinnabar (mercury). A mineral composed of lighter elements can also have a high refractive index if the atoms are closely packed and held together by strong bonds. The outstanding example is diamond, which is composed of the light element carbon, but in which the atoms are so strongly held together that the mineral has a very high refractive index and a high luster.

Hardness

The resistance a mineral offers to scratching is a measure of its hardness. Like refractive index, this is related to the crystal structure, for the more tightly its atoms are bonded together, the more difficult it is to scratch a mineral. Diamond is the hardest of all substances and nothing will scratch it except another diamond. At the other end of the hardness spectrum are minerals like talc and graphite, so soft that they have a smooth, slippery feel, and rubbing them between the fingers will tear off sheets of atoms. Although this property is measured qualitatively it is easy to determine and is thus important in mineral identification.

The German mineralogist, Friedrich Mohs (1773–1839), arranged ten relatively common minerals in the order of increasing hardness and since that time the hardness of all minerals has been rated according to his scale.

Mohs Scale of Hardness

1. Talc
2. Gypsum
3. Calcite
4. Fluorite
5. Apatite
6. Feldspar
7. Quartz
8. Topaz
9. Corundum
10. Diamond

If an unknown mineral can be scratched by a mineral in the scale, it obviously has a lesser hardness; if it cannot be scratched, it has a greater hardness. The property is thus determined by trial and error; for example, a mineral is rated 5½ if it can be scratched by feldspar but not by apatite. The steel of a knife blade has a hardness of 5½, a copper coin about three, and a fingernail a little over two.

Cleavage

The electrical forces acting between atoms frequently vary with the crystallographic direction. A crystal may therefore break easily parallel to a set of atomic planes that are at right angles to the direction of the weakest binding force. Such easy breaking of crystals, yielding plane surfaces, is called cleavage. Since cleavage is controlled by internal structure and takes place only between atomic planes, it is always parallel to crystal faces or possible crystal faces. It is therefore expressed in terms of its crystallographic direction; for example, cubic cleavage indicates three directions (one parallel to each *pair* of parallel cube faces) along which the crystal will break with equal ease. The mineral halite, common salt, illustrates this type of cleavage. If a few grains of table salt are examined with a magnifying glass it will be seen that each grain is bounded by six plane surfaces at right angles to each other as a result of its cubic cleavage.

Some minerals have such good cleavage that they can be broken only parallel to these definite crystallographic directions. But in others, cleavage is much less obvious and is developed with difficulty. Cleavage is thus measured by the ease with which it can be accomplished and the quality of the surface produced as well as crystallographic direction. For example, mica has one direction along which it can be easily split into exceedingly thin sheets; whereas the single cleavage in beryl is difficult and imperfect. Some minerals have several cleavages but others none at all. Thus, the number of cleavage surfaces, their quality and crystallographic directions are important properties in mineral identification.

Specific Gravity

We all have some concept of how much an object of a given size and appearance should weigh. We expect a chunk of black iron to be heavier than a light-colored stream pebble of the same size, for we know in general that metallic objects are heavier than nonmetallic. This comparative weight, known as specific gravity,[2] is one of the easily determined and important determining properties of minerals. It is given as a number expressing the ratio of the weight of a mineral to the weight of an equal volume of water. Thus if a mineral specimen weighs four pounds, and an equal volume of water weighs one pound, the specific gravity is four.

The bulk of the rocks that we most frequently encounter are made up principally of three minerals, feldspar, quartz and calcite, and their specific gravities are 2.60–2.75, 2.65 and 2.72 respectively. Their average specific gravity is thus probably between 2.65 and 2.75. To a lesser extent we also have a sense of the specific gravity of metallic substances. Graphite (sp.g. 2.2) seems light, whereas a block of silver (sp.g. 10.5) of the same size seems heavy. The specific gravity of pyrite (5.0) is usually considered average for metallic minerals.

To determine specific gravity we must know the weight of the mineral and of an equal volume of water. The mineral can of course be weighed very precisely. To determine the weight of an equal volume of water we make use of the well-known principle: a body immersed in water is buoyed up and weighs less than it does in air, and the loss of weight equals the weight of the water displaced. The mineral fragment is first weighed in air and is then weighed again while suspended in water by a fine thread. The difference between these two weights is equal to the weight of water displaced. Expressed as a formula:

$$\frac{\text{Wt. in air}}{\text{Wt. in air} - \text{Wt. in water}} = \text{specific gravity}$$

For example, a quartz crystal weighs 4.55 grams in air and 2.83 grams when suspended in water; the specific gravity is therefore:

$$\frac{4.55}{4.55 - 2.83} = \frac{4.55}{1.72} = 2.65$$

[2] Specific gravity and density are here used interchangeably. However, density should be stated in units of measure, as, for example, grams per cubic centimeter or pounds per cubic foot.

(Above) Orange-red needles of crocoite crisscross cavities in brown to black limonite from Dundas, Tasmania. (Katherine H. Jensen)

(Below) Orpiment from Mercer, Utah, shows the sulfur-yellow color that is its most distinguishing characteristic. (Harry Groom)

The density of a mineral, like its luster and refractive index, depends on the atoms of which it is composed and the manner in which they are packed together. Thus minerals with a high luster usually have a high density. It follows also that a mineral with a high percentage of atoms of a heavy element such as tin, tungsten, lead or mercury will have a high density. Conversely, minerals composed essentially of light atoms usually have low density. However, exceptions to this rule are brought about by the influence on specific gravity of such other factors as atomic arrangement and interatomic forces. If atoms are held together by strong bonds, a much larger number can be packed into a given volume than if the bonds are weak. This is illustrated most dramatically by diamond and graphite, both composed of the light element, carbon. The close packing of atoms in diamond, reflecting the powerful forces holding them together, gives it a specific gravity of 3.5, high for a nonmetallic mineral. On the other hand, in graphite the atoms, held loosely by weak forces, are comparatively far apart, giving it the much lower specific gravity of 2.2.

The two most fundamental properties of a mineral, its chemical composition and crystal structure, can be determined only by an expert using special equipment. Fortunately for the less skilled person, these properties are reflected in the easily observed or measured characteristics of crystal form, color, luster, hardness, cleavage and specific gravity. These properties are sufficiently diagnostic or indicative to serve for the identification of most minerals. Appendix I lists the minerals mentioned in this book with their crystallographic and physical properties.

Crystals of pyrite (from Elba, Italy), the most common sulfide mineral, show a variety of crystal forms, but all, including the two cut stones, have a brass-yellow color. (Studio Hartmann)

2 Early Use of Minerals

Recent discoveries in Africa have revealed that mankind has made tools of stone for nearly two million years. Prior to these finds, it was thought that toolmaking was a relatively late chapter in the human story. Now it is clear that man's ancestors shaped their destiny in stone, and by their success in the use of tools determined the course of their biological evolution.

The Toolmakers of Olduvai Gorge

When in 1953 L. S. B. Leakey and his wife began their work in Olduvai Gorge in what is now the Republic of Tanzania, they found evidence that the region had been the dwelling place of tool-using animals for a very long time. Associated with the remains of animals long extinct were a great many crude "pebble tools," some of which had been modified slightly by chipping or flaking. Found in cavelike rock shelters, at a distance from the creek bed where they occurred, there was evidence that they had been brought to the caves for a purpose. These artifacts of a prehuman stone industry are collectively known as the "Oldowan culture." Who used these simple but tremendously significant tools was not at first known, but between 1955 and 1963 the Leakeys uncovered skeletal remains of fourteen different hominid individuals in the ancient stream deposits and the beds of volcanic ash in which the Olduvai Gorge is carved. Some of these individuals are members of a genus called *Australopithecus,* meaning southern ape, which has no living descendants. The others belong, as does modern man, to the genus *Homo,* and are in the direct line of our ancestry. These are presumably the toolmakers of Olduvai Gorge, and the species to which they belong has been fittingly called *Homo habilis,* meaning able, handy, skillful, vigorous. The beds in which the men of Olduvai

(Above) Flint nodules from France (gray) and Egypt (brown) with artifacts made from similar materials about forty thousand years ago. (Emil Javorsky)

(Below) Beautifully formed flint arrow heads found in the Sahara Desert attest to the skill of paleolithic craftsmen. (Pierre A. Pittet)

Chinese porcelains of the K'ang Hsi period of the Ch'ing Dynasty (1662 to 1722) show the mastery of the Chinese potter. Both the saucer (above left), 8½ inches in diameter, and the plate (below), 24⅛ inches in diameter, are in the Metropolitan Museum of Art. (The Metropolitan Museum of Art, Bequest of John D. Rockefeller, Jr., 1960)

(Above right) Pre-Columbian gold figure from Chimu, Peru, is set with irregularly shaped stones of turquoise. (Pierre A. Pittet)

Gorge left their bones and tools are, as dated by the radioactive potassium-argon method, 1.65 to 1.75 million years old! A new mineralogical method of age determination, based on tracks left by radioactive particles in minerals, confirms this, placing the age of the oldest rocks in the gorge at about two million years.

Of the close relation between early man and minerals Sherwood L. Washburn, professor of anthropology at the University of California, says: "Now it appears that man-apes ... had already learned to make and to use tools. It follows that the structure of modern man must be the result of the change in the terms of natural selection that came with the tool-using way of life." In this view, the achievement that set the man-ape on the road to civilization was a mineralogical one, the selection of suitable materials for tools. A pebble can be thrown with deadly effect; a stone lends killing force to the blow of a fist, but if broken so as to yield a sharp edge, it can be used to trim fresh, strong sticks to usable shapes. The transition from tool-using to toolmaking involved the appreciation of the property of quartz called *conchoidal fracture*. If a pebble of crystalline quartz or flint is struck, either by design or chance, against another pebble, it may break in a curving, shell-like fracture. Two such fracture surfaces, meeting along an edge, yield a razor-sharp cutting tool. Obsidian, or natural volcanic glass, shares this property with the various forms of quartz, and hence, like quartz, is a good tool material. Most other common minerals do not break in this way and also lack the hardness and toughness of quartz. It reflects favorably on the practical sagacity of the man of Olduvai that over a period of nearly two million years, even where supplies were remote and difficult to obtain, quartz, flint and obsidian remained almost exclusively the materials for weapons and tools. Until the beginning of the use of metals, about 5500 to 6000 years ago, there was little change in the materials sought for tools. Indeed, no better natural materials could have been chosen. Dr. Leakey, using an ancient quartz tool, has reportedly demonstrated to scoffers his ability to skin a freshly killed gazelle as rapidly and effectively as a competitor using a steel knife.

There is nothing in a quartz pebble or flint concretion that suggests the functions of the knife, axe or scraper that it may become. The fashioning of a useful tool from such a pebble can be the product only of both accidental observation and purposeful experiment. The end result is an appreciation of some of the outstanding properties of the mineral quartz: its hardness and its conchoidal fracture.

Edward S. Deevey, Jr., director of the Geochronometric Laboratory at Yale University, estimates the entire human population of the earth one million years ago at perhaps 125,000 individuals. Little increase in total population took place during the long span of the Old Stone Age, and he estimates a population of only about five million during the flowering of Paleolithic culture that produced the cave paintings of the Pyrenees and the beautiful "laurel leaf" flint points of Solutré, France. Although this total human population seems small, Deevey estimates about thirty-six billion Paleolithic hunters and gatherers lived during the course of the Old Stone Age. Considering that in an average life span of about thirty years, a man of the Old Stone Age used thousands of flint tools, it is not surprising that worked flints from that epoch are still common finds. It is more surprising that after so long a period of dependence on

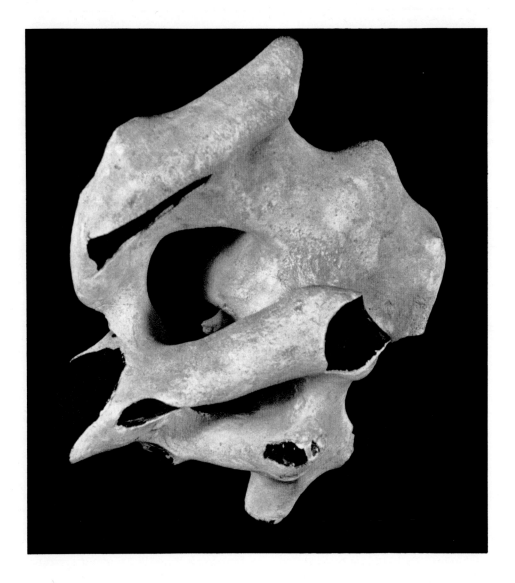

From the celebrated Chalk Cliffs of Dover of southeastern England come flint nodules in many shapes and sizes. This specimen is white because of the chalk but dark along the fresh fracture. (Cornelius S. Hurlbut, Jr.)

stone tools civilized man should have completely forgotten them. By classical Greek and Roman times, they were thought to have fallen from the sky as the result of a bolt of lightning. This strange belief persisted up into the nineteenth century in some parts of the world.

The realization that the many shaped flints found in Europe were simply the discarded tools and weapons of prehistoric Europeans finally came in the eighteenth century. This resulted in part from the growing knowledge of the culture of American Indians, who were still practicing the arts of the Stone Age. However, the art of knapping or flaking flint never wholly disappeared in Europe. Flints were used as fire-making tools in flint-and-steel kits, and in the era of flintlock firearms large quantities of them were needed. During the American Revolution and the Napoleonic Wars, gunflints were as essential as gunpowder. Gunflints are still knapped in the vicinity of Brandon, England, for export to primitive areas where flintlocks are still used and for the convenience of collectors of antique guns. Such gunflints are simply wedge-shaped tablets and lack the finesse of the best Stone Age weapons.

Shaping Flint

Flint, coarsely crystalline quartz or even common bottle glass may be knapped in two ways: by percussion or by pressure. In percussion flaking, a flint is struck with another piece of flint, knocking off a bulbous flake; it is thus gradually shaped into a "core" which may then be refined into an axe or chopping tool by further flaking, called retouching. Alternatively, the detached flakes may be treated as tools and finished by retouching. In either the "core" or the "flake" tradition, the final edging and finishing is done by means of pressure flaking, in which a bone, wooden or horn tool is pressed with a wrenching twist against the edge of the object being retouched. Small flakes are twisted away alternately from upper and lower surfaces, leaving a somewhat serrate but effective cutting edge. A skilled worker can prepare an "Indian arrowhead" from a fragment of bottle glass in five or ten minutes.

A deeper insight into the relationship between early man and his most prized mineral resource comes from the excavations in the chalk downs of southern and eastern England where Neolithic miners dug for flint. Using deer-antler picks, and shoulder-blade-bone shovels, these expert miners sank hundreds of pits in the chalk as much as fifty feet deep in search of beds yielding the best flint nodules. Lateral workings were driven from the bottom of the pits along the flint-bearing beds. Laboring in low, cramped galleries by the light of saucer-shaped grease lamps fabricated from chalk, the miners extracted the vital nodules, which were then fashioned into tools. Crudely carved figurines of pregnant women and other fertility symbols have been found in the flint pits along with offerings of antler picks. Apparently the miners thought of the flints as the result of some generative force in the rocks, a force which required placation.

Egyptian fresco from the time of King Thutmoses III depicts slave labor breaking ore-bearing rocks.

The art of extracting metals from ores is very old: shown here are stages in the smelting of gold-bearing ore by ancient Egyptians. (Historical Pictures Service-Chicago)

Primitive Clay Modelers

The use of tools began very early and it was a long time before tool-using man learned to use clay and became pottery-making man. With the exception of flint, clay is the mineral that has been longest and most closely associated with man's destiny. The fabric of human history is picked out with gay, brilliant weft threads of gold, silver, and copper, strengthened with the somber hue of iron, and embellished with diamond, ruby and emerald; but through it all run the dull, unspectacular warp threads of clay.

What child has not felt the elemental fascination of clay? That smooth, wonderfully plastic material responds to every changing fancy. Lucky is the child who discovers a natural bank of clay and having modeled some objects and forgotten them returns to find that the sun has converted the clay into a brittle but firm stone.

There is a significant difference between mud and clay, a distinction very important to human history. Both may be semiliquid, but when clay is molded and dried it retains its molded shape and has a firmness that potters call "leather-hard"; whereas when mud dries, it merely crumbles to dust. The leather-hard clay object can be trimmed with a knife, decorated by impression or incision, or even turned, like wood, in a lathe. However, even when thoroughly dried and seemingly quite hard, it remains clay, and if wetted, slumps at once into its original semiliquid state. Primitive men, discovering the possibilities of the leather-hard state, modeled figurines, vessels, and grotesque images that needed only formalization to become gods. A few of these creations have survived in the caves and grottos of the Pyrenees and of Australasia.

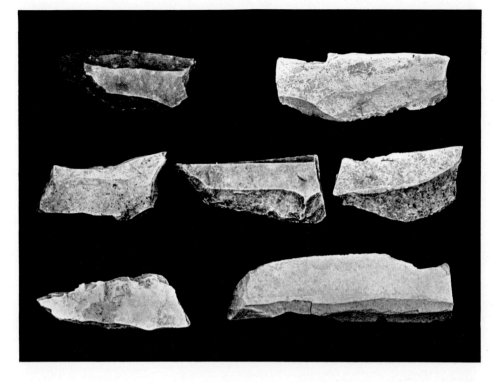

Because of the sharp edge created at the intersection of two fractures, flint was the most important raw material of early "toolmaking man." Although primitive, these flint scrapers from Dordogne, France, have plainly been formed by the directed breaking of flint nodules. (Courtesy of the American Museum of Natural History)

Bison and canoes modeled from the clay of a cave floor have endured, although unbaked and unfired, for tens of thousands of years.

Clay was used in other ways. Smeared over the inside of a rush or reed basket, it could make the basket sufficiently waterproof for carrying water short distances. Pressed as a glob over the lid of a storage pit and marked with a thumb print, it served as a seal of ownership on the grain within.

The First Firing of Clay

At last there came the marriage of clay and fire, perhaps accidentally, that gave man the boon of pottery. When a clay object is buried in a bed of coals so that the clay is brought to glowing red heat, a magical change takes place. The leather-hard clay, which would slump if wetted, changes to a rocklike material that is no longer affected by moisture. Although porous, the fired clay retains its shape and is endowed with a truly remarkable strength. Pots, bowls, storage jars and lamps can all be made easily and quickly. Patterns pressed into the soft molded pieces will be retained in the rock-hard product of firing. Potters soon learned to glaze their pots, coating them with a "slip" or paint of watery clay, which when fired formed a glassy crust impervious to water. The use of colored glazes, with or without incised or raised decoration, gave rise to changing styles that can be used to date pottery. Since nothing is more durable than a piece of broken pottery, a potsherd, it is on pottery styles that archaeologists have based the chronology of prehistory. The trained archaeologist can pull out a few broken bits of pottery from a mound and date it at once as "Early Islamic," "Fourth Century Roman," "Bavarian," etc. Constructing a pottery chronology is complex, but the principles are simple. The introduction of new

Late Neolithic pottery attests to the technological skill and artistry of the potter. (Courtesy of the American Museum of Natural History)

designs and techniques from outside the area resulted in the increasingly rapid obsolescence of styles. Thus in the ruins of ancient civilizations the archeologist finds layers of pottery that reveal the succession of periods.

The Magic of the Potter's Kiln

What is the mineralogy behind the magic of the potter's kiln? "Clay" is a loose generic term for a whole group of related minerals, all characterized by a sheetlike crystal structure in which the elements oxygen, silicon and aluminum predominate, and in which there is some water. In chemical terms the clay minerals are essentially hydroxyl-bearing alumino-silicates with a sheet structure. The simplest in composition and structure is kaolinite. When pure, kaolinite occurs as snowy-white compact masses and loose aggregates of submicroscopic flakes. Each flake is a crystal made up of layers bound together by very weak electric forces. And each layer consists of oxygen, hydrogen, aluminum and silicon atoms strongly bound together into the sheetlike structure.

When mixed with water, the submicroscopic flakes, lubricated by water molecules, glide readily over one another, giving the clay plasticity. When dried, the flakes again cohere, yielding the leather-hard conditions; however, the internal crystal structure of the flakes remains the same. When fired red hot or somewhat hotter, a profound change takes place: the hydrogen and some of the oxygen in the kaolinite escapes as water, leaving behind the aluminum silicon-oxygen framework. The sheetlike arrangement remains but the crystal structure is permanently changed into a product called "metakaolin" which, if wetted, will not revert to liquid clay. However, the ceramic resulting from a firing at low temperature is rather soft. Some very primitive pots fired in open camp fires are of this kind and are called "low-fired earthenware." If the temperature is raised to white heat (over 1000° C or 1800° F) the sheetlike structure collapses and a much harder and less porous product called "stoneware" results. If firing is continued to even higher temperatures, say 3000° F, the grains of mineral matter partially melt and fuse together into a kind of semiglassy state. The ceramic ware so produced is very hard, completely nonporous, rings when struck, and in thin pieces is usually translucent. This ware is called "porcelain" or "fire china" after the country where it was first made.

The Pottery Industry

Although porcelain was first made in China, European potters eventually learned the secret of high-fired translucent porcelain. Staffordshire, England, the Limoges district in France, and the region around Dresden in Saxony have long been famous as centers of manufacture of fine porcelains. In France, in the sixteenth century, Bernard Palissy, a self-taught genius, searching for the secret of Chinese porcelain, produced a rustic pottery of great charm which found favor with the French nobility and established pottery manufacture in France. In England in the eighteenth century, Josiah Wedgewood made earthenware and stoneware that was both beautiful and practical. His kilns in Staffordshire formed the nucleus of the modern English ceramic industry.

North and south of Zanesville, Ohio, in the United States, is a topographic feature known as "Potter's Ridge." Here good potter's clay crops out in close association with bituminous coal. Because the clay occurs below the coal in the sequence usual in sedimentary beds, some of the richest and most often exploited beds are called "underclay." Along Potter's Ridge during the nineteenth century rose the smoke of hundreds of primitive pottery kilns. So numerous were potters, who made chiefly coarse but serviceable articles, that a descendant says, "One cannot sink a spade anywhere on Potter's Ridge without turning up fragments of fired pottery." The pottery is called "stoneware" because the high temperature of firing produced a strongly bonded material that looks like gray stone on a fracture surface. These wares were heavy, hard and well suited both to the needs of a rough frontier and later to the chores on the farms of the area.

The local clay was mixed with "opening material," generally crushed fragments of fired pottery, called "grog," but sometimes with sand or shale included. Mixed with water, the clay-grog mixture was kneaded to expel the occluded air. It was then turned by hand on the potter's wheel into heavy bowls, crocks and jugs, often with pleasing forms. These were allowed to dry thoroughly and then stacked in a small kiln separated by pieces of fired clay called "kiln furniture." A blue decoration was frequently applied on the dry but unfired ware, using a mixture of cobalt oxide, "smalt," clay and water, mixed to the consistency of paint. The fire was then built up in and around the kiln, using wood, charcoal or, later, coal. At the height of the firing process, common rock salt, thrown into the kiln, instantly attacked the surface layer of the fired ware, causing it to melt and form a glassy glaze. The usual product of this process was a dense, hard gray pottery with a rather uneven glaze, called "salt-glazed stoneware."

A fascinating look into the past may be had by visiting any of the numerous potteries near Uhrichsville, Ohio. Here large sections of clay pipe used for sewers are made by firing in "beehive" kilns that must look much like those used in ancient Samaria and Babylonia. The firing process is, moreover, identical except that the fuel may be gas or oil rather than charcoal or wood.

Between the time when the flint miners worked on the chalk downs, perhaps 3000 B.C., and the Early Middle Ages, there were no true developments in the thinking about minerals. A considerable body of practical knowledge had accumulated, but it was neither systematized nor founded on any consistent theories of the origin or nature of minerals.

Minerals in Ancient Egypt

In Egypt of the predynastic period (5000–3000 B.C.) the list of natural minerals in use was already long. Native metallic copper was used in tools and weapons. Gold and silver were plentiful. Agate, carnelian, chalcedony, chrysoprase, jasper and rock crystal, all varieties of quartz, were employed as ornaments. Turquoise, olivine, chrysocolla, green feldspar (amazonite), jade, green fluorite and malachite, all found in predynastic settings, attest to the Egyptian's love for the color green. Malachite, a green copper mineral, was extensively used as an eye paint, and an elaborate ritual involving palettes for its preparation and utensils for its application seems to have existed before dynastic times. Galena, garnet, hematite, and lapis lazuli are also known in the form of beads, seals and pigments. At later periods of Egyptian history this list lengthened greatly. The numerous minerals were valued partly for their beauty but even more for their magical powers, and were clearly the object of a prolonged search. Most of these minerals are not common, and few occur in the inhabited part of the Nile Valley, so the prospectors who sought them went far afield. These hardy adventurers of antiquity must have had some sort of systematic way of classifying minerals and metals and they were probably expert in reading surface signs that pointed to valuable minerals within an outcrop. They also must have had a system of tests for the identification of the minerals they sought. They probably possessed as wide a knowledge of geography and as much information about the properties of materials as any men of their time. Yet because of their humble birth and the manual, hence lowly, character of their occupation they remained uncelebrated. Thus their knowledge, unrecorded, disappeared when they died. In the "Report of the XV International Geological Congress," 1929, modern geologists in reporting on the gold deposits of Egypt remark that the ancient Egyptians "were very thorough prospectors and no workable deposits have been discovered that they overlooked"—a remarkable tribute.

3 The Earth and Its Rocks

Minerals are the building units of rock. Thus, for an understanding of the mode of origin and distribution of different rock types, one must have a knowledge of the physical and chemical properties of minerals. Conversely, in order to have a full appreciation of the origin, association and occurrence of minerals, one should know something about the rocks in which they are found. Consequently, it is impossible to wholly divorce minerals from rocks; and the present chapter summarizes the major features of the three rock types: the igneous, the sedimentary and the metamorphic.

The earth that most of us take for granted as a symbol of permanence and stability did, as *Genesis* says, have a beginning, and its origin has been for a long time the subject of discussion by cosmogonists, astrophysicists and geologists. Of the several theories of origin, most agree that the material of which the earth and the other planets is made had a common origin with the sun. In some hypotheses, the material constituting the planets is considered to have separated from the sun and later aggregated; in some, sun and planets are supposed to have originated together by aggregation of dispersed material.

A valid hypothesis of earth origin would be easier to formulate if we knew the nature of the material beneath the earth's skin. The study of earthquake waves has given us considerable information about the interior of the earth and shows it to be composed of concentric layers separated by discontinuities. Earthquake waves, caused by sudden fracturing of the brittle crust, move out in all directions from the source and, like light waves, change their direction when they enter a medium of different density, or are reflected from the interface between two such media. From the interpretation of seismograms, or earthquake records, a major change in earth material is found to take place at a depth of eighteen hundred miles below the surface. Below this discontinuity is the core, about 4400 miles in diameter, believed to be composed essentially of iron and nickel. The central part of the core reacts to earthquake waves as a solid, whereas the outer twelve hundred miles reacts as a liquid.

Another marked change in the rocks is found much nearer the earth's surface. It was first noted by Mohorovičič, a Yugoslav scientist, and is thus called the Mohorovičič discontinuity but frequently referred to merely as the "moho." Beneath the ocean basins it lies from five to eight miles below sea level, whereas beneath the continents it is usually about twenty-one miles below sea level but extends as deep as thirty-five miles under high mountain ranges. Above this rather irregular surface is the crust, and extending downward from it to the core is a large zone called the mantle. Both crust and mantle are composed essentially of rocky material. There is general agreement that there is a threefold division of the earth into crust, mantle and central core.

All evidence points to the fact that the heaviest material is at the center of the earth in the iron-nickel core, the lightest in the crust, and material of intermediate density in the mantle. Thus, whatever theory of origin one accepts, it seems necessary to postulate that during its formative period the earth was molten throughout. Only if such a state existed is it possible to account for the increasing density from surface to center. In a liquid state, under the effect of gravity, the heavier materials would tend to sink toward the center, displacing the lighter materials upward.

When geologists speak of rocks, it is the stony material of the crust to which they refer, for the studies of the geologist have been confined largely to the rocks he can reach with his hammer on the earth's surface itself. What lies below remains in large part a matter of inference and speculation. Our knowledge is slightly increased by the examination of the materials brought up from deep mines and still deeper wells. In the Witwatersrand conglomerate at Johannesburg, South Africa, miners have descended to a depth of two miles in search of gold. In Pecos County, Texas, the world's deepest oil well has pierced limestone, sandstones and shales to a depth of nearly five miles. Although these are monumental achievements for mining and petroleum engineers, they contribute very little to our basic understanding of the constitution of the earth, for both the mines of the Witwatersrand and the deep oil wells of Texas bottom in rocks that are in no way different from those at the surface. We know that the rocks explored in this way were at one time laid down at the earth's surface on yet older rocks.

If samples of the material from the mantle could be examined, their study would shed light on our understanding of the earth's interior that until now has been based on conjecture. Preliminary steps of the ambitious project of drilling a hole through the crust to the mantle beneath the moho was underwritten by the U.S. National Science Foundation. This undertaking, called Operation Mohole, has been abandoned for lack of funds. However, fragmentary accounts from the Soviet Union indicate that the Russians are drilling holes from which they may soon bring samples of the mantle rock to the surface.

If we assume that throughout its long history the earth has maintained its layered structure, the elements now in the upper part of the crust were there from the beginning. Countless rearrangements of them have been taking place at the surface and for a few miles beneath, but the bulk composition has remained nearly constant. In the following table are given the percentages of the twenty most abundant elements of the earth's crust.

Shown here are various shapes and colors of mica: (above) long brown crystals of phlogopite mica from Ontario, Canada, embedded in white calcite; (center above) diamond-shaped crystals of white mica, muscovite, from Ontario; (center below) a small crystal of muscovite surrounded by a rim of lithia mica, lepidolite; and (below) a large fragment of purple lepidolite made up of an aggregate of minute scales. (Emil Javorsky)

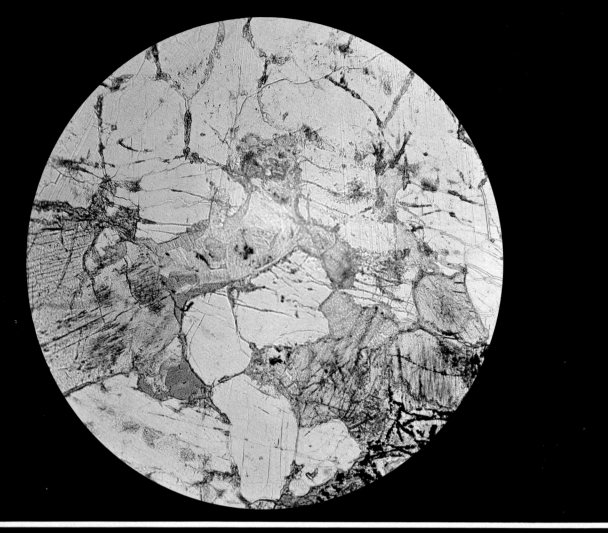
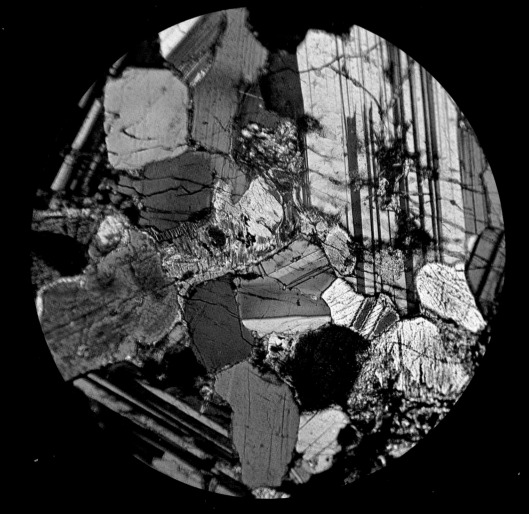

Abundance of the Chemical Elements in the Earth's Crust in Weight per cent[1]

Element	%	Element	%
Oxygen	46.6	Phosphorus	0.12
Silicon	27.72	Manganese	0.10
Aluminum	8.13	Fluorine	0.07
Iron	5.00	Sulfur	0.05
Calcium	3.63	Strontium	0.05
Sodium	2.83	Barium	0.04
Potassium	2.59	Carbon	0.03
Magnesium	2.09	Chlorine	0.02
Titanium	0.44	Chromium	0.02
Hydrogen	0.14	Zirconium	0.02

[1] This table is based on the averages of many individual analyses.

Although ninety-two elements are found in nature, the eight most abundant constitute nearly ninety-nine per cent of the total amount. It is from these, then, that minerals of the rocks of the crust were first made and are made today. The dominant element is oxygen, not a gas as in the air we breathe, but as charged atoms bound to metal atoms by powerful electrical forces. Of the metal atoms, silicon is most important, for the rock-making minerals are silicates, that is, compounds containing oxygen-silicon groups. The common mineral quartz is composed of only these two elements.

As shown in the table, oxygen is vastly more abundant than any other element, and, furthermore, it is light, occupying more space than the heavier metallic elements. Thus if we consider percentage by volume instead of by weight, oxygen makes up nearly ninety-four per cent of the solid rocks. The crust can then be pictured as a boxwork of oxygen atoms joined together by silicon and the other common metal atoms.

Although the list of rock-making minerals is long, only a few make up the bulk of the rocks. Further, there are many interesting varieties of the common minerals that are considered elsewhere; only those important to the understanding of the rocks are listed below.

A thin section of a course-grained rock of granitic texture observed in ordinary light (above) and between crossed polarizers (below), magnified twenty-five times. The colors of the mineral grains of plagioclase feldspar, mica, hornblende and augite, almost indistinguishable in ordinary light, can easily be seen under the polarized light. (Tozier Collection, Harvard University)

The Major Rock-Making Minerals

		Color	Cleavage	Hardness	Specific Gravity
Dark	Olivine	Olive-green	None	6½–7	3.27–4.37
	Pyroxene (Augite)	Dark-green to black	2 at about 90°	5–6	3.3
	Amphibole (Hornblende)	Dark-green to black	2 at about 125°	5–6	3.2
	Mica				
	Biotite	Black	1 excellent	2½–3	2.8–3.2
Light	Muscovite	White	1 excellent	2½–3	2.76–3.1
	Feldspar	White to gray	2 at 90° or about 90°	6	2.57–2.76
	Quartz	Colorless	None	7	2.65

Igneous or Fiery Rocks

If we assume our primitive earth was a fiery-hot globe of molten material, the first solids formed were at the surface as heat gradually escaped into outer space. This initial thin layer was an igneous rock and the material from which it solidified was a magma. The name igneous (from the Latin for fire) is applied to any rock that is formed from a cooling melt. Thus in the beginning all rocks were igneous, and the other types we now know were derived from them.

Intrusive, Extrusive and Pyroclastic Rocks

To the west of the Bitterroot range that marks the Continental Divide in northwestern United States lies the wilderness country of the Salmon River Mountains. It is an area of rugged beauty, much of it accessible only by foot or pack animals. Thus the traveler has ample time to consider each rock exposure as he comes to it. He finds after a few miles that each outcrop bears a monotonous likeness to the one preceeding—a gray rock speckled with flakes of black mica. Nor does it change for many more miles, for the Salmon River Mountains are carved from the Idaho batholith, a gigantic body of uniform rock that covers sixteen thousand square miles, twice the area of Massachusetts. When the sun shines, flashes of light are reflected from thousands of tiny crystals at every outcrop. Close examination shows that the flashes come from a gray mineral that makes up most of the rock and imparts its color to it. The mineral is feldspar and the reflecting surfaces are its two cleavages. The rock is a granite made up dominantly of feldspar, but glassy grains of quartz and scattered flakes of the black mica, biotite, are also present. These crystals range in size from one-eighth to one-quarter of an inch, forming an interlocking pattern called "granitic texture."

The Idaho batholith is an example of one group of igneous rocks, the intrusive, formed by the intrusion of magma from below into the upper layers of the earth's crust, but still at considerable depth below the surface, where it crystallized slowly to form a coarse-grained rock. After millions of years, erosion has removed the overlying rocks, exposing at the surface the intrusive rock formed far below.

As one travels west across the Salmon River Mountains the terrain becomes more subdued and gives way, before reaching the Oregon boundary, to flat or rolling country. A marked change in the rock accompanies this change in topography. The gray rock of the mountains with the flashing crystals is replaced by a black massive rock most of which shows no identifiable minerals. It is basalt, a type of extrusive igneous rock formed from a magma that has poured out on the earth's surface as a lava flow. Similar rocks known as the Columbia Plateau basalts are found in eastern Oregon and Washington over an area of nearly 200,000 square miles.

Compared with intrusive rocks that have had time to crystallize slowly to form coarse textures, extrusive rocks have so fine a texture as a result of rapid crystallization that the individual grains cannot be distinguished with the unaided eye. In fact some lavas have cooled so rapidly that no crystals have formed and the resulting rock is a glass. If a magma began to crystallize in depth before it was extruded as a lava flow, the relatively large early-formed

A thin section of basaltic lava observed in ordinary light (above) and between crossed polarizers (below), magnified twenty-five times. Large phenocrysts of olivine (left) and augite (right) are enclosed in a fine-grained matrix composed essentially of plagioclase feldspar. The colors in the lower photograph result from the interference of the polarized light. (Tozier Collection, Harvard University)

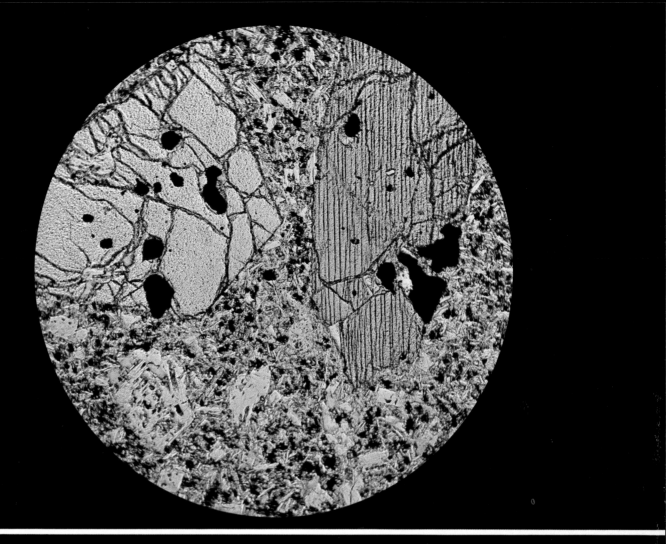

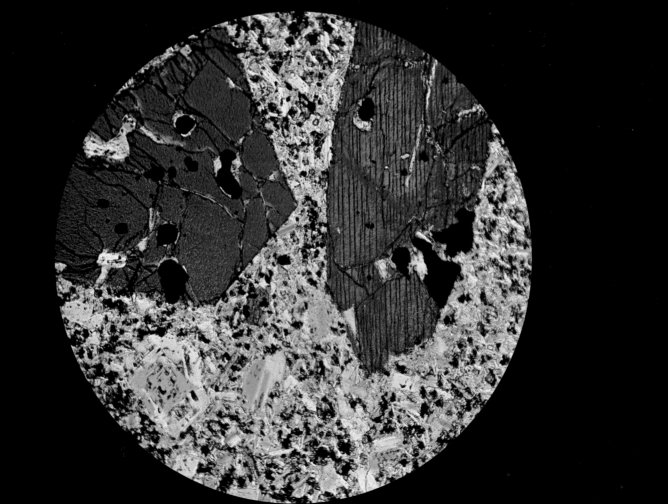

crystals will be found frozen into a later-formed fine-grained matrix. These crystals are called phenocrysts and the rock has a porphyritic texture.

An igneous rock has essentially the composition of the melt from which it formed, but water and other volatiles that may comprise several per cent of the magma escape during crystallization. When a magma that has been confined under high pressure suddenly reaches the surface, these volatiles form bubbles that rise through the lava and escape. But some become trapped by advancing crystallization and remain, forming gas cavities in the flow. Such open spaces may become the sites of later deposition of a great variety of minerals.

The great expanse of the Columbia Plateau is made up of many essentially horizontal lava flows, but interlayed with them are pyroclastic rocks. These are of similar chemical and mineralogical composition to the associated basaltic lava flows, but are composed of broken fragments of rock crystals and volcanic glass. Most lava flows issue quietly from fissures in the earth's crust, but at times the gases confined in the magma escape with such violence when the magma reaches the surface that rock material is blown into the air. Fine material is carried by the wind and may settle over a large area as a thin bedded deposit of volcanic ash, forming on later consolidation a rock called tuff. Coarser material does not travel as far and is found near the site of the explosive activity as volcanic breccia.

A magma of a given composition may form intrusive, extrusive or pyroclastic rocks that are mineralogically similar and differ only in texture and method of formation. But mineralogically and chemically different types of intrusive rocks and their volcanic equivalents occur, so magmas of different chemical compositions must exist. Igneous rocks are named largely on the kinds and amounts of minerals present that reflect the chemical composition of the magma from which the rock crystallized. Thus a magma high in iron forms a rock containing iron-rich minerals while one with abundant calcium forms calcium-rich minerals. However, when a large magma reservoir has been emplaced in the earth's crust and conditions permit it to crystallize slowly, several different rock types may result by a process called magmatic differentiation. As the magma slowly cools, minerals begin to separate in a sequence determined by their solubilities, the least soluble crystallizes first. Since there is plenty of time, these early-formed minerals react with the cooling magma to produce other minerals. If at any point in this process a geological disturbance separates the liquid part from the crystallized solid, the solid remains unchanged. The liquid part, on the other hand, may begin a new series of precipitations and reactions to form a rock quite different from the earlier separated solid portion.

Other inhomogeneties may arise in a cooling magma even without intervening geological accidents. The earliest minerals to come out of the molten solution are the dark ones—first olivine, followed by pyroxene and amphibole. All of these minerals have specific gravities greater than that of the melt from which they separated. They may thus settle before any reaction can take place and build up in successive layers at the bottom of the magma chamber. All three of these dark minerals are rich in iron and magnesium, and thus the still fluid upper part of the magma becomes impoverished in those elements but at the same time enriched in aluminum, sodium and potassium, and may

(Above) Polished sodalite with white areas of nepheline, from Bancroft, Ontario, Canada. (Katherine H. Jensen)

(Below) A large yellow-brown crystal of idocrase shows the square cross section characteristic of tetragonal symmetry. (Studio Hartmann)

Feldspar crystals with square cross sections protrude into cavities in a granite at Baveno, Italy. What appears to be a single crystal is a "baveno twin," two crystals grown together according to a definite law. (Alfred Ehrhardt)

solidify to form a rock such as granite. There can then result, by this process of magmatic differentiation, a series of rocks with different chemical and mineralogical composition.

The Kinds of Igneous Rocks

The diversity of igneous rocks is so great that over six hundred names have been used to indicate different types. Most of these names involve subtle distinctions of importance only to the specialist, or pertain to rare rocks. Only a few names are necessary to indicate the broad groups into which most of the igneous rocks fall.

If rock classifications were based entirely on the kind and amount of dark mineral present—olivine, pyroxene, amphibole, and biotite—naming of a rock would not be difficult. These minerals can be recognized and their percentages estimated with comparative ease. However, most classification schemes depend on the amount and kind of feldspar, because feldspar is almost always present. Feldspar is an alumino-silicate containing potassium, sodium and calcium. One of these elements is generally dominant, although most feldspars contain all three in varying amount. To name a rock with certainty it is necessary to know not only how much feldspar is present but its chemical composition as well. This is an impossibility without x-ray or microscopic study, for all feldspars look much alike in the small grains of the average intrusive rock. However, the specific gravity varies with changes in chemical composition and can be interpreted in terms of the chemistry.

In general the feldspars are divided into two groups: (1) the potash feldspars, containing potassium, and (2) the plagioclase feldspars, containing sodium and calcium. As shown in the table below, the plagioclase feldspars form a complete series between albite (ab), the pure sodium member, and anorthite (an), the pure calcium member. The naming of the intermediate members is quite arbitrary and their compositional range is indicated by the percentage of *ab* and *an* they contain.

The Feldspars

	Name	Chemical Composition	Specific Gravity
Potash feldspar	Orthoclase (monoclinic)	$KAlSi_3O_8$	2.57
	Microcline (triclinic)	$KAlSi_3O_8$	2.57
Plagioclase feldspar	Albite (ab)	$NaAlSi_3O_8$– ab 90	2.62
	Oligoclase	ab 90–70;	
	Andesine	ab 70–50;	INCREASING
	Labradorite	ab 50–30;	
	Bytownite	ab 30–10;	
	Anorthite (an)	ab 10– $CaAl_2Si_2O_8$	2.76

The intrusive igneous rock found most extensively at the earth's surface is granite. It forms the core of many of the great mountain ranges and can be seen today in lofty peaks of the Rocky Mountains, the Alps and the Himalayas. Because of their resistance to erosion, granite masses in many areas stand above the general level of the country, forming rounded domes such as the "Sugar Loaf" at Rio de Janeiro and Stone Mountain in Georgia. If we could strip off the comparatively thin veneer of surface rocks on the continents we would find granite world-wide, forming the principal rock of the upper part of the earth's crust.

Granite is a light-colored rock composed of feldspar and quartz with small amounts (about ten per cent) of other minerals, chiefly mica and hornblende. The mica may be either muscovite or biotite or both. In addition, tiny scattered grains of accessory minerals, such as magnetite, ilmenite, apatite, zircon and sphene may be present. In some granites the feldspar is almost exclusively a potash variety, however a typical granite contains about thirty per cent each of potash feldspar, plagioclase (oligoclase) feldspar and quartz. With an increasing percentage of plagioclase the rock grades into a granodiorite. The color of a granite is determined by the feldspar, which may be colorless, various shades of gray, buff or even red. The quartz is usually in colorless, glassy grains.

The volcanic or extrusive equivalent of a granite is a rhyolite, containing the same minerals in the same proportions but with so fine a texture that one cannot recognize the individual mineral grains. Some streaks through the rock may be noncrystalline glass, but if the rock is composed entirely of glass it is called obsidian. Black obsidian is known in many parts of the world, the most famous occurrence being Obsidian Cliff in Yellowstone National Park. It was used extensively by the American Indians for arrow heads and it is used today in Mexico as cut stones for jewelry. Pumice is a glassy rhyolite in which gases escaping from the magma as it reached the surface produced a frothy texture. Cavities in it are so numerous that it is extremely light in weight. In certain areas it is not uncommon, following a torrential rain, to see boulders of pumice floating downstream. After the eruption of the volcano Krakatoa in 1883, floating pumice blocks were a hazard to navigation on the Indian Ocean.

A rock much less common than granite but resembling it in general appearance and texture is syenite. In mineral composition syenites are also similar to granite except they lack quartz. They are made up, therefore, essentially of a mixture of potash and plagioclase feldspar with some dark minerals, hornblende and biotite mica. Small amounts of apatite, sphene, zircon and magnetite are present as accessory minerals. If the plagioclase feldspar exceeds the potash feldspar, the rock is called a monzonite. In general a monzonite is darker than a syenite, for an increase in the dark minerals usually accompanies an increase in plagioclase. The volcanic equivalent of syenite is the uncommon light-colored rock trachyte. The bulk of the rock is made up of a fine-grained, even-textured groundmass through which may be scattered recognizable crystals of feldspar and hornblende.

The intrusive rocks diorite and gabbro are much less common than their extrusive counterparts, andesite and basalt. They are dark rocks lacking quartz and potash feldspar and are characterized by the plagioclase feldspar: oligoclase-andesine in diorite and labradorite-anorthite in gabbro. However, a distinction can usually be made on the basis of the dark minerals: hornblende in diorite, and pyroxene with some olivine in gabbro.

Magnetite, ilmenite and apatite are the common accessory minerals usually present in diorite and gabbro in the form of scattered grains. But in some gabbro bodies segregations of magnetite and ilmenite may be sufficiently abundant, as at Kiruna, Sweden, and Sanford Lake, New York, to form giant ore bodies.

Andesite, the extrusive equivalent of diorite, is a comparatively rare rock except in one region in the world. This is in the Andes Mountains of South America where andesite is a dominant rock. The rock and the chief feldspar in it, andesine, are both named for this major occurrence.

Basalt, the chemical and mineralogical equivalent of a gabbro, is a black, dense and fine-grained rock. It is the most common and abundant of the volcanic rocks; in fact more basalt is now at the earth's surface than all of the other extrusive rocks combined. Since the beginning of geologic time, basalt flows have periodically poured forth from fissures. These lavas may in places be many thousands of feet in thickness and cover hundreds of thousands of

A fragment broken from a lava flow at Mt. Vesuvius contains abundant, nearly spherical-shaped crystals of leucite. (Emil Javorsky)

square miles. In eastern Washington and Oregon and southern Idaho, the Columbia Plateau, covering an area of 200,000 square miles, was formed by the outpouring of basaltic lavas four to six thousand feet thick. In central India vast floods of basalt, the Deccan traps, covering an area of half a million square miles, poured out in successive flows one upon the other. Since each flow was somewhat less extensive than the preceding, they formed giant steps that one must ascend to reach the plateau. The name trap, taken from the Swedish word *trapp* meaning step, was first applied to these basaltic rocks of India. Since then the term *traprock* has been used to refer to any of the fine-grained dark rocks whose mineral composition may be uncertain.

Elsewhere in the world less fluid basaltic magmas have piled up as volcanos. The great mass of rock forming the Hawaiian Islands is basalt, built into volcanic peaks reaching from the sea floor fifteen thousand feet below sea level to an elevation of fourteen thousand feet above sea level. The numerous "cinder" cones that dot the southwestern United States and Mexico are, like the volcanic islands of the South Seas, constructed of basalt, but are fragments or scoria rather than flows.

Feldspar, so important in most igneous rocks, may be present in only small amounts or completely absent from some. These include the peridotites, composed essentially of the dark minerals, pyroxene and olivine. If pyroxene is the dominant mineral, the rock is a pyroxenite. When olivine is dominant, the rock is a dunite. Some dunites, as in the Dun Mountains of New Zealand, from which locality the rock receives its name, are composed almost exclusively of olivine.

The great variety of igneous rocks results from the different proportions of the chemical elements in the magmas from which they crystallized. Some

*The horizontal striations on this cleavage fragment of plagioclase feldspar from Perth, Ontario, Canada, are the result of repeated twinning.
(Elmer B. Rowley)*

magmas are so rich in silica that they form rocks containing quartz, pure silica uncombined with other elements. Other magmas have just enough silica to form silicates, chiefly feldspar, but with none left over to form quartz. Syenite is an example of a rock crystallizing from such a magma. There are still other magmas with insufficient silica to form the feldspars of normal igneous rocks but which do form other interesting minerals, the feldspathoids. Chief among these are nepheline and leucite, but sodalite, lazurite, haüynite and noselite are also formed. The number of different rock types containing feldspathoids is large, but most common are nepheline syenite and leucite syenite in which these minerals take the place of some of the feldspar of ordinary syenite. The largest mass of such rocks is on the Kola Peninsula in northwest Russia. Here a circular area covering about six hundred square miles is composed of a variety of nepheline-bearing rocks with which are associated many rare and unusual minerals.

The lavas that form the great mass of Mount Vesuvius and the volcanic ash ejected in the eruption that overwhelmed Pompeii in A.D. 79 contain feldspathoids. They are fine-grained dark rocks that look much like basalt but contain an abundance of multi-faced, nearly spherical crystals of leucite.

Sandstone and Other Sedimentary Rocks

The weathering of any rock, igneous, sedimentary or metamorphic, produces fragmented or dissolved material, most of which ultimately becomes incorporated in a sedimentary rock. One broad division of sedimentary rocks, the detrital or fragmental, includes those composed of (1) pieces broken mechanically from earlier rock, (2) mineral particles liberated from source rocks, and (3) new mineral particles formed by chemical reactions during the weathering process. It is only after the detrital particles in any of the types become firmly cemented or compacted that the body is considered a *rock*.

Mechanical weathering usually produces angular rock fragments that, if transported only a short distance from their source, preserve their angularity. If these angular fragments become cemented together the resulting rock is called a *breccia* (pronounced bret'sha).

Most breccias result from catastrophic processes such as landslides, cave collapse or volcanic eruption. Far more commonly, the fragments are transported long distances by streams, waves or moving ice before they reach their final resting place. In this process they become smooth and rounded as they rub against one another and against the bedrock over which they move. An accumulation of such rounded boulders, cobbles and pebbles after cementation is called conglomerate, or, quaintly, pudding stone. Oliver Wendell Holmes furnished a sprightly but unscientific explanation of the Roxbury conglomerate, a well-known formation in the Boston area, in the last three stanzas of his poem "The Dorchester Giant":

> They flung it over to Roxbury hills
>   They flung it over the plain,
> And all over Milton and Dorchester too
> Great lumps of pudding the giants threw;
>   They tumbled as thick as rain.
>
> Giant and mammoth have passed away,
>   For ages have floated by
> The suet is hard as a marrow bone,
> And every plum is as hard as stone,
>   But there the puddings lie.
>
> And if, some pleasant afternoon,
>   You'll ask me out to ride
> The whole of the story I will tell
> And you shall see where the puddings fell,
>   And pay for the punch beside.

Many minerals resist chemical attack during the weathering process and are removed intact as tiny grains from the enclosing rock. Of these, quartz alone is in significant amount and is the major and sometimes the only mineral constituent in the sands of the streams, the deserts and the sea beaches. Sand grains of quartz caught up by the wind and drifted into dunes, as in the Sahara Desert, accumulate to form large deposits on land. More common and of greater magnitude however, although unseen, are the accumulations under water, particularly in the sea.

Small quartz grains forming such deposits are transported by stream currents sometimes thousands of miles before being swept out to settle in the comparative quiet of ocean water. The small rounded sand grains spread out layer upon layer on land or on the ocean floor become sandstone, but only after they are cemented firmly together. It is chemically precipitated mineral matter that locks the sand grains firmly in place and is largely responsible for the color of the rock. Most quartz is either white or colorless, thus if white or gray calcium carbonate, the commonest cement, surrounds the grains, a white or gray rock results. In many sandstones the iron oxides, goethite and hematite, nature's most common pigmenting materials, stain the calcareous cement or act as cementing material themselves. The yellows and brown result from the mineral goethite and the reds from the mineral hematite.

Some sandstones result from the accumulation of river-borne deposits in the valley and on the flood plain of a stream and are therefore called fluviatile sandstones; others form in the deltas of great alluviating rivers where these empty into the sea and are called deltaic sandstones. Sandstones have a granular, sugary appearance, and although firm and compact, crumble to sand grains if vigorously rubbed or pounded. This differentiates them from quartzites, which, like sandstones, are often wholly made up of quartz grains, but are so firmly cemented together that they break through the quartz grains rather than around them and hence do not yield sand upon grinding or pounding.

In the weathering of rocks, some minerals are chemically unstable and break down to form new compounds, some soluble in water, others insoluble. The insoluble portion of the products of chemical weathering is mostly clay, composed of countless submicroscopic, flaky particles. When these are washed into streams, they are kept in suspension even in slowly moving water and hence are carried to the sea. When the streams with their load of sediment reach the sea, the coarser sandy material first settles out, forming sandstones, then granular particles so small that the naked eye cannot distinguish the individual grains. These are called siltstones. The flocculant clay particles are carried farther than the coarser sand, mud and silt and settle in the deeper portions, which on consolidation become shale. The chief characteristic of shale is a pronounced tendency to split into thin layers and flakes, giving the weathered outcrop the appearance of an untidy stack of papers.

Another class of sedimentary rocks comprises chemical compounds precipitated from solution in the water of the sea and of lakes. Their formation may be caused by chemical reaction, evaporation, or by the activities of living organisms. These rocks are also formed from pre-existing rocks, but in this case the earlier rock was reduced by weathering to atoms or ions in solution rather than broken into fragments mechanically. Most abundant in this class of chemically precipitated rocks is limestone. Some limestones are practically pure calcite, made up of chemically precipitated calcium carbonate, or by the accumulation of shells. Others contain considerable amounts of clay, sand, dolomite, or organic matter. Also included in this category of sedimentary rocks are those deposited by evaporation of sea water: salt, gypsum, anhydrite, magnesite and sylvite; and those deposited by evaporation of desert lakes. These economically important minerals will be treated separately.

Metamorphic Rocks

Although rocks are sometimes taken as a symbol of permanence and stability, they undergo significant changes not only by weathering at the earth's surface but also far below the zone of weathering. It is this latter type of change, in which the rock-forming minerals are recrystallized essentially in the solid

The cooling of basaltic lava that once poured over the coast of County Antrim, Ireland, created tension cracks, giving rise to these hexagonal-shaped columns known as the Giant's Causeway. (Boris Heuss: Bavaria-Verlag)

state, that is called metamorphism. If a volcanic or sedimentary rock becomes deeply buried, it is subjected to high pressure and temperature, conditions quite different from those under which it formed. Minerals that were stable at the time of formation become unstable in the new environment and the atoms of which they are composed rearrange themselves. New minerals may form and old minerals may disappear, or by recrystallization give rise to new structures and textures. During the process of metamorphism sometimes nothing is added or subtracted, but small amounts of chemically active fluids present in the rocks aid in solution and redeposition.

If metamorphism takes place in an environment of high confining pressure, there is a tendency for minerals to form that bring about a reduction in volume. Thus many of the stable new minerals are frequently those in which the atoms are packed closely together and as a consequence have a high specific gravity. Many of these minerals rarely occur under other conditions and are characteristic of metamorphic rocks. However, feldspar and quartz, so abundant in other rock types, are also common in metamorphic rocks.

Gneiss and Schist

The feature that characterizes most metamorphic rocks is a banded or layered structure called foliation. The individual bands or folia range in thickness from a small fraction of an inch to eight or ten inches and differ from one another in color and texture, reflecting a difference in mineral content. Coarsely foliated rocks are called gneiss (pronounced *nice*); whereas the more finely laminated rock is called schist.

The banding of a gneiss results from segregation of light-colored minerals, chiefly quartz and feldspar, into layers alternating with layers rich in dark minerals. Although gneisses are commonly derived from sedimentary rocks, they often resemble an igneous rock and are given names such as granite gneiss or syenite gneiss. If a mineral such as hornblende or biotite is particularly abundant, the rock may be named hornblende gneiss or biotite gneiss, a better practice because it does not involve genetic assumptions.

In contrast to coarsely banded gneisses, schists possess schistosity along which they may be easily broken. This results from an abundance of platy minerals, such as micas, chlorite and talc, oriented with the plane of their excellent cleavage parallel to the schistosity. These minerals have developed during metamorphism with their broad dimensions at right angles to the direction of greatest pressure. The easy splitting of schists results from the parallel arrangement of the flaky minerals and their good cleavage.

In addition to the platy minerals, schists frequently contain equidimensional or sharply-angular minerals that have crystallized with characteristic forms in spite of the directional pressure under which they developed. These include staurolite, sillimanite, kyanite, andalusite and certain varieties of garnet. A rock with a given bulk chemical composition subjected to relatively low pressure-low temperature conditions will develop a characteristic set of minerals. But as the intensity of metamorphism increases, there is a corresponding mineralogical change in both the platy and sharply-angular minerals. Thus, in the metamorphism of a shale new minerals develop as temperature and pressure increase, permitting the rocks to be grouped into zones characterized

(Above) At Twin Falls, Idaho, the Snake River has cut a valley through basaltic lava flows piled one upon the other. (Mary Shaub)

(Below) Sheer walls of varicolored sandstone rise above the Virgin River in Zion National Park in southwestern Utah. (Benjamin M. Shaub)

by certain critical minerals. In order of increasing intensity they are: chlorite, biotite, garnet, staurolite, kyanite and sillimanite. Schists are sometimes named after the dominant mineral, as mica schist, chlorite schist; or after a conspicuous mineral, as garnet schist, staurolite schist, kyanite schist.

Marble, Quartzite and Serpentine

In igneous rocks and in shales several minerals are present and their metamorphism gives rise to a variety of rocks. If, however, a rock is composed essentially of one mineral, as in the sedimentary rocks limestone and sandstone, there is little opportunity to develop new minerals. Nevertheless, recrystallization takes place when these rocks are subjected to metamorphic conditions.

Marble is a metamorphic limestone which, when pure, is a snowy-white rock made up entirely of the mineral calcite. The sedimentary textures of the limestone completely disappear during metamorphism and the fine-grained calcite reforms into interlocking grains. This recrystallization may produce in some marble coarse grains in which the calcite cleavage is easily seen, but in others the individual grains can be distinguished only with a magnifying glass. However, only rarely is a limestone pure calcite. It may contain interbedded layers of clay or other minerals which during metamorphism develop a flow pattern characteristic of many decorative marbles. In Burma, rubies, a variety of corundum, have formed from aluminum-rich clay impurities during metamorphism of limestone to marble. In a recrystallized limestone in New Jersey, called the Franklin limestone (really a marble), the initial scattered impurities have been concentrated in newly formed crystals, and isolated crystals of spinel, chondrodite, pyrrhotite, graphite and phlogopite mica are found.

Dolomite is also the major, if not the only mineral, in some marbles. Such dolomitic marbles have been derived from sedimentary rocks in which the mineral dolomite is the principal mineral. Although somewhat harder, they resemble calcite marble so closely that without a chemical test it is difficult to distinguish between them.

Quartzite is a metamorphic sandstone. Under great pressures the rounded quartz grains lose their spherical shape. Material is dissolved from a grain at the points of greatest pressure to be reprecipitated at points of lesser pressure giving rise to an interlocking texture of irregularly shaped grains. If the sandstone contains no impurities, the resulting quartzite is white. But if an impurity such as iron oxide is present, the quartzite may be brown, pink or even deep red.

Quartzites may form also from sandstone at relatively low temperatures and pressure through the medium of circulating solutions containing silica. Each sand grain, originally broken from a quartz crystal, acts as a nucleus or "seed" to control the crystallographic orientation of the new silica deposited from solution. Since new quartz is added simultaneously to all grains, each crystal can increase in size only slightly before encountering its enlarging neighbors. If the process continues until all the original cementing material has been replaced and all the voids filled, the rock is made up of an interlocking aggregate of quartz grains. Microscopic examination of a thin section of such a rock may show outlines of the original rounded sand grains indicated by specks of

(Above) Transparent blades of mottled blue kyanite in a matrix of fine-grained mica from St. Gotthard, Switzerland.
(Studio Hartmann)

(Below left) Finely crystallized specimen of epidote from Knappenwand, Austria.
(Emil Javorsky)

(Below right) Although diopside usually occurs in small grains it is sometimes found in transparent crystals of sufficient size and quality to be cut into gem stones.
(Benjamin M. Shaub)

Nearly hemispherical domes of granite rise above the general level of the terrain near Bulawayo, Rhodesia. (Cornelius S. Hurlbut, Jr.)

iron oxide, but crystallographic continuity is maintained in the newly added quartz.

In some sandstones the above process has added insufficient silica to each sand grain to fill the voids and lock them firmly together. Although the grains cannot be separated, a slight movement is permitted between them. Such a rock usually contains muscovite and is called itacolumite or flexible sandstone. If a slab of this rock fifteen inches long and half an inch thick is laid on a table, the center can be raised as much as an inch without causing the ends to leave the table.

During the metamorphism of peridotites, there is usually a complete mineralogical change from the pyroxene and olivine of the original rock. If water is present, these magnesium and iron rich minerals are altered to serpentine, a soft ($H = 2-4$) waxlike mineral with a greasy luster. This mineral is the principal constituent of serpentine rock, which is usually massive and varies in color from light to dark green. When cut and polished, serpentine rock may show beautiful variegated coloring, and is thus used extensively, under the name verde antique marble, as an interior decorative stone.

Minerals From Contact Metamorphism

When an igneous magma makes its way from the depths into the upper portion of the earth's crust, the rocks which it encounters may undergo changes. The alterations that occur at or near the contact, known as contact metamorphism, are evidenced by the formation of new minerals in the invaded rock. In some cases recrystallization is brought about merely through the increase in temperature; in others, chemically active solutions from the magma may produce significant mineralogical changes in the country rock. Usually, the larger the body of intruded magma, the greater the zone of contact metamorphism.

The type and extent of the metamorphic change depends on the nature of the magma but even more on the country rock. Thus, sandstone is converted to quartzite whereas shale becomes "hornfels," a dense rock composed of a fine-grained mixture of several minerals. The most striking changes are brought about when a granite magma intrudes limestones and dolomites. The heat of the intrusion alone may cause recrystallization to a marble, and various impurities may crystallize into such minerals as spinel, corundum, graphite and wollastonite. As limestone and dolomite react chemically with solutions from the magma, even more profound changes take place, giving rise to a large and characteristic group of minerals. The calcium of the limestone and the calcium and magnesium of the dolomite combine with silica supplied by the magma to produce an abundance of calcium and magnesium silicate minerals, most typically grossularite and andradite garnet, idocrase, epidote, tremolite and diopside. Many other minerals may also be formed, including ores of copper, lead and zinc. Thus, in some areas contact metamorphic deposits are of great economic importance as a source of metals.

4 Minor Minerals of the Rocks

The great variety of igneous rocks that form on crystallization of molten magma results largely from different combinations of a relatively small number of common rock-forming minerals. In addition, there are usually present a small percentage of other constituents known as accessory minerals, commonly in the form of tiny, well-formed crystals. In column one below are listed those minerals whose main occurrence is as accessories in igneous rocks. In column two are those accessories found abundantly in other geologic environments and considered elsewhere in this book.

Accessory Minerals of Igneous Rocks

1	2
Zircon	Magnetite
Rutile	Ilmenite
Monazite	Hematite
Sphene	Pyrite
Apatite	Garnet

Of the five minerals in column one, zircon, rutile and monazite are resistant to chemical attack and have a high specific gravity. Thus, as the rocks containing them are decomposed by weathering they are liberated to find their way intact into the streams and eventually to the sea. Washed by coastal currents, they may be carried hundreds of miles and end as placer deposits in beach sands. Such concentrations exist along the eastern coast of Florida. Here, associated with ilmenite, have been mined thousands of tons of tiny crystals of zircon, rutile and monazite that originated in the Appalachian Mountains far to the north. Similar concentrations are found in many beaches and coastal

(Above) A flashing blue brilliant of synthetic rutile (left) bears little resemblance to the large reddish natural crystal (bottom) of the same mineral. Zircon (second from left) is seen in a cut stone, variety hyacinth, with a small brown crystal below. The wedge-shaped crystals and yellow cut stones are sphene. (Studio Hartmann)

(Below) Yellowish crystals of gem sphene and white crystals of feldspar coat a cavity at Tavetsch, Switzerland. (Katherine H. Jensen)

dunes, with the largest in India, Brazil and Australia. Although the major accumulations of these three minerals are confined to this type of deposit, isolated occurrences of larger crystals are of more interest to the mineralogist.

Zircon and Other Accessory Minerals

For centuries zircon has been recovered from the stream gravels of Ceylon and Burma and used as a gem. It is usually a shade of red or brown but more rarely may be green, violet or even colorless. The red and the yellow, known as hyacinth and jacinth respectively, have long been highly prized as gem stones.

When liberated from the enclosing rock, zircon is always found in the form of square prismatic crystals terminated by a pyramid. Although its hardness (7.5) is greater than that of quartz, the crystals recovered from stream gravels are usually somewhat rounded and the faces indistinct. But a water-worn pebble when cut has a brilliance that approaches that of a diamond, and the colorless zircons from Matura, Ceylon, were called "Matura diamonds." Even today many a colorless zircon is set in an engagement ring for the young man unable to buy the far more expensive diamond.

The most popular zircon is the beautiful blue stone, with a touch of green, sold under the name *starlite*. Although some crystals of zircon are naturally pale blue, a far more attractive color is produced by heating. Most of the gem zircons that reach western markets have been heat-treated. When heated in air, brownish, off-color stones usually turn a golden yellow but some become colorless; if heated without air, a blue or colorless stone may result. The art of producing more attractive gems by heating was a secret passed down from one generation to another, and only recently has the process been scientifically recorded.

Zircon is a zirconium silicate ($ZrSiO_4$) in which the German chemist, M. H. Klaproth, in 1789 discovered the then new element zirconium. Since then it has been learned that most zircon contains about two per cent of hafnium, an element first discovered in zircon in 1923. Thus the tiny crystals of zircon accumulated in beach sands are the ores of two metals used in the construction of nuclear reactors. However, most of the zircon because of its chemical inertness and high melting temperature is used in the manufacture of refractories and ceramics.

Rutile, another accessory mineral, occurs in crystals very similar in shape to zircon and is usually found in slender square prisms terminated by a pyramid. But it lacks the glamour of the zircon for, although it has a high luster, rutile found in nature is rarely used as a gem. The mineral is a dark red to black and the occasional stone that is cut has a high metallic luster like gun metal. Long slender needles of rutile are present in rock crystal known as rutilated quartz, or Venus' hairstone.

With advancing technology many minerals, including rutile, have been synthesized. It is made by the Verneuil process, the same method used for the synthesis of ruby and sapphire. Natural rutile is essentially titanium dioxide, TiO_2, but usually contains minor impurity elements. By using chemically pure starting material, single crystals of rutile have been produced that are nearly colorless and have only a faint yellow tinge. Cut into stones, this synthetic material with refractive index and dispersion higher than diamond makes

Apatite of gem quality has been found in many places; notable occurrences are the richly colored purple crystals from Auburn, Maine (above left), and the yellow-green crystals from Durango, Mexico (above right). (Both by Benjamin M. Shaub)

*(Below) Asparagus stone, gem quality apatite crystals, grown into a cavity lined with quartz and feldspar.
(Katherine H. Jensen)*

Crystals of wavellite radiating from a common center form spherical aggregates in a quartzite from Montgomery County, Arkansas. (Emil Javorsky)

dazzling gems. Their hardness is unfortunately low and they are subject to scratching when worn in jewelry.

The rutile that is recovered from beach sands is used as an ore of titanium metal. The largest use of this element is as a paint pigment in the form of the oxide, but rutile is too impure and the pigment is obtained principally from the mineral ilmenite.

The mineral monazite acquired its name, which comes from the Greek word *monazein*, meaning to be solitary, because it occurs in isolated grains. In fact, it is only through the agency of running water that, with one exception, significant accumulations have been formed. The exception is in Cape Province, Republic of South Africa, where a dikelike body of monazite has been mined.

Monazite is a phosphate mineral, $(Ce, La, Y, Th)PO_4$, containing several of the so-called rare earth elements, including thorium, and it is because of the thorium that monazite commands scientific and commercial attention. Late in the nineteenth century thorium nitrate was used in the manufacture of the popular Welsbach gas mantle. Although thorium has some minor uses today, its major use in the future will undoubtedly be as a source of atomic energy, for it is needed in the preparation of uranium, U^{233}, a fissionable isotope.

Although sphene is an abundant accessory mineral in many igneous rocks, its low hardness and easy cleavage prevent its accumulation in placer deposits and beach sands. Interest in it is centered on those few localities where it is found in rocks in large amounts or in crystals of exceptionally fine quality. Large crystals of sphene occur in many colors but the pale yellow and green transparent varieties are of most interest to the gemologist. Stones cut from such crystals have a brilliant luster with a fire surpassing that of diamond and make gems of incomparable beauty. However, they lack an essential attribute of a gem stone—hardness. They are much too soft to be used in jewelry. The principal source of transparent sphene crystals is the St. Gotthard district in Switzerland. Another source but of non-gem material is on the Kola Peninsula of northwestern Russia where a group of nepheline syenite rocks containing concentrations of many unusual minerals is found. Sphene is scattered through these igneous rocks; but at one locality it is found as a major constituent. Since 1932 in this area, called the Khibina tundra, the Russians have been mining sphene as a source of titanium.

Apatite is a mineral of low hardness and high solubility. Therefore, unlike many of the other accessory minerals, it does not accumulate in placers or beach sands, but disintegrates in place, as rocks are weathered. It is fortunate that it does so for apatite is calcium phosphate, $Ca_5(PO_4)_3F$, and although when it breaks down chemically some of the phosphorus is washed into the streams and eventually to the sea, some remains behind in the soil and provides one of the elements most important for plant life.

The phosphorus taken up by plants is passed on to the animals, including man, that feed on vegetable life and is incorporated in their bone structure. Our teeth are phosphates with a chemical composition very similar to apatite. Some of the phosphorus that reaches the ocean forms chemical compounds that accumulate on the sea floor; some is taken up by fishes and other marine life and when they die their hard parts settle to the bottom and may build up thick beds of phosphates. It is from such ancient deposits, now elevated above sea level, that we derive the phosphate for fertilizers necessary for agriculture. Millions of tons of this phosphate are mined each year. The world's largest deposits are in northern Africa, and all of the countries bordering on the Mediterranean produce it. In the United States, it is produced principally in Florida, Tennessee, Montana and Idaho.

In Quebec and Ontario in Canada, well-formed crystals of apatite are found in coarse crystalline limestone in such large concentrations as to permit commercial mining of phosphate. On the Kola Peninsula in Russia, there are even larger deposits of well-crystallized apatite, associated with nepheline syenite rocks. Since 1932 these have supplied enough phosphorus for the fertilizer needed in the vast agricultural programs of the Soviet Union.

Apatite is usually nontransparent and some shade of green or brown, but it

A well-formed crystal of augite (far right), one of the dark minerals of igneous rocks, is embedded in a fragment of fine-grained rock. Because of the resistant nature of staurolite (right), crystals frequently weather intact from the metamorphic rock in which they grew. (Studio Hartmann)

is also sometimes found in yellow, blue, violet, or colorless transparent crystals that may be cut into gems. Apatite of gem quality has been found in many places; two of the most notable occurrences are the richly colored purple crystals from Auburn, Maine, and the yellow-green crystals of Durango, Mexico. Magnificent stones have been cut from them but unfortunately the low hardness makes apatite a poor gem.

Metamorphic Minerals

The common minerals in igneous rocks—feldspar, quartz, mica and hornblende—are also common in metamorphic rocks. But metamorphic rocks also contain minerals that have resulted from their formation under high temperature and high pressure. These metamorphic minerals, although in rocks of similar chemistry, are not always the same, for those formed at a given temperature and pressure may differ from those formed under other conditions. The most interesting of these minerals are garnet, staurolite, kyanite and andalusite; they are frequently found in large crystals distributed through the rocks, each with its characteristic shapes.

Garnet, the most widespread of these minerals, is frequently found in nearly spherical red crystals scattered through a schist or gneiss like plums in a pudding. This mineral is found in many varieties and geologic environments and is considered in a later chapter under colored stones.

Staurolite is found in well-formed crystals with characteristic shapes. Some crystals occur singly but more interesting are the "twins" which are composed of two crystals interpenetrating one another and forming a cross, with the arms of the cross at an angle either of about 60° or 90°. These cruciform twins, particularly the right-angle type, have long attracted attention and, without cutting or polishing are worn in Christian countries as pendants. They are usually nontransparent and of a reddish-brown to dark-brown color. But on Mount Campione and St. Gotthard in Switzerland lustrous translucent crystals are associated with gem quality kyanite.

Staurolite is resistant to chemicals and since it is harder than most of the minerals with which it is associated, it remains behind in the soil as the enclosing rock is disintegrated by weathering. In the mountainous country near Taos, New Mexico, crystals can be picked up on the surface over a large area underlain by schist. Farmers also turned up these staurolite crosses in the piedmont region of Virginia and North Carolina and under the name of "fairystones" or "fairy crosses" they were then sold to tourists as amulets. The crop of crystals is smaller each year and to satisfy the demand enterprising natives carve "fairystones" from a soft rock, color them to the red-brown of staurolite and market them in gift shops. The man-made cross can be easily distinguished from the natural one; it is easily scratched by a knife whereas true staurolite is too hard to be scratched. The fraud can also be detected by measuring the angles between the crystal faces, for a carver cannot duplicate the precision of the interfacial angles reflecting the arrangement of atoms in the true crystal.

The Sillimanite Group

The minerals, andalusite, sillimanite and kyanite are polymorphs of the same

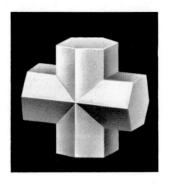

More common than single crystals of staurolite are twins composed of two intergrown crystals. Two types of twins are equally common—one with the individuals interpenetrating at a 90° angle (above), and the other at an angle of about 60° (below).

Crystals of kyanite from Phoenix, Arizona, growing from two directions meet to form a spear-shaped aggregate. (Floyd R. Getsinger)

chemical compound; that is, although chemically identical, each has its own crystal structure and related physical properties. Although the rocks in which they are found may have originated from the same material, each mineral represents a different set of pressure and temperature conditions of metamorphism. Commercially these minerals are called the "Sillimanite Group" although sillimanite itself is the least important and has very little industrial value. In the many places where they make up ten to forty per cent of the rock, andalusite and kyanite have been mined as the raw material for refractories and high-grade porcelains.

Sillimanite is a comparatively rare mineral but it is of interest to the petrographer in that it is formed under the highest temperature conditions. It is usually distributed throughout rocks in long slender colorless crystals called fibrolite. Occasionally pale blue transparent crystals of sillimanite are found in the stream gravels of Burma and Ceylon and are cut as gem stones.

For thirty years following its discovery in the Inyo Mountains of California, andalusite was valuable in the manufacture of spark plugs. In this inaccessible area, with transportation only by burro, the mineral was mined near the craggy summit of White Mountain at an elevation of ten thousand feet. Although more ore remains, the operation was abandoned when it was discovered that equally good or better spark plugs could be made with aluminum oxide. Andalusite is still being mined in the Transvaal of South Africa; in Switzerland and in Kazakhstan, Russia.

Andalusite occurs in elongated crystals with a square cross section which shows, in the variety known as *chiastolite*, a black cross formed by the crystallographic arrangement of carbonaceous inclusions. Some peoples believe that such crystals have magic powers and sections cut from them are frequently worn as amulets. The common colors of andalusite are reddish-brown or green and the rare transparent crystals that are cut into gem stones may show a strong dichroism, appearing green when viewed from one direction and red from another.

Kyanite characteristically occurs in bladed crystals that may be white, gray or green but are most typically blue with the darker shades toward the center. Many minerals possess a different hardness in different crystallographic directions but none so striking as kyanite: crystals of kyanite can be easily scratched with a knife parallel to their length but cannot be scratched at right angles to this direction. Transparent varieties of kyanite are occasionally cut as gem stones. As a commercial refractory mineral, kyanite is the most abundant and important member of the "sillimanite group" and India is the chief producer, the principal deposit being at Lapsa Buru in the northern part of the state of Orissa. In the United States the most important of the many workable deposits are in Virginia, North and South Carolina, and Georgia.

(Above) Nearly black almandite garnets in a mica schist from Monroe, Connecticut, all exhibit a trapezohedral crystal form.
(Katherine H. Jensen)

(Below left) Garnet crystals, each surrounded by a rim of black hornblende, like huge plums in a giant pudding, occur at Gore Mountain, New York, where garnet has been mined as an abrasive since 1880. (Cornelius S. Hurlbut, Jr.)

(Below right) Emerald-green uvarovite garnet crystals with the rose-red chromium-bearing chlorite, known as kotschubeite.
(Harry Groom)

5 Nature's Treasure House

Three examples of green microcline feldspar, called amazonite: cleavage fragment and necklace from Amelia Courthouse, Virginia, and crystal from Pikes Peak, Colorado. (Emil Javorsky)

The industrial city of Córdoba in central Argentina lies in the foothills of Serra de Córdoba, a mountain range, composed of north-south ridges of pre-Cambrian granite and gneiss. When one looks from the first ridge west of the city across a valley to the crest of the second, the most conspicuous feature on the skyline is a gleaming white peak, appropriately called Cerro Blanco. The peak stands above the surrounding rocks because of the great resistance to weathering and erosion of the coarse milky-white quartz of which it is largely composed. Associated with the quartz are large blocks of equally white feldspar and small amounts of muscovite mica. Cerro Blanco is a pegmatite and attracts the mineral collector—not for its quartz, feldspar and mica but for the uncommon and rare minerals associated with them.

From the geological point of view, pegmatites are rather small bodies of rock rarely more than a mile in maximum dimension. Their mineralogical composition commonly resembles that of a granite, but they differ in having an extremely coarse texture and containing large crystals intergrown in an irregular fabric. In addition to containing the common igneous rock minerals, they are the home of hundreds of rarer species that may replace the more abundant minerals or fill fissures and cavities. They are commercially important both as the source of many gem minerals and of numerous raw materials used for industrial purposes. Furthermore, they present a fascinating problem in earth history, since their mode of origin is not fully understood. For these reasons pegmatites are well worth detailed consideration.

Pegmatite Areas

Pegmatites occur in abundance everywhere that crystalline rocks are exposed by erosion. Thus, in the great pre-Cambrian areas of the continents such as the

Canadian Shield, where millions of square miles of granite and granitic gneiss are exposed, pegmatites are found in great numbers. They also appear in smaller areas where local or regional doming or mountain uplift has raised ancient crystalline rocks to be uncovered by erosion. Thus, in the Black Hills of South Dakota, where a blister-like bulge has brought the underlying granitic rocks to the surface, some of the most famous pegmatites in North America occur. Other famous pegmatite areas of the United States are in the metamorphic terrains of New England, the Blue Ridge, the Rockies of Colorado and New Mexico, and the igneous rocks of southern California. Pegmatites are found in the ancient rocks of every continent, and especially in Southwest Africa, Mozambique, Southern Rhodesia and Tanzania in Africa and Madagascar; Sweden, Finland, Norway and Germany in Europe; Russia and India in Asia; Australia; Canada and Mexico in North America; Brazil, which affords the world's most notable pegmatites, Argentina and Bolivia in South America.

An unusual and somewhat mysterious feature of pegmatites is the fact that wherever mining, quarrying or natural erosion has fully revealed the contacts, they appear to be wholly enclosed by other rocks. The avenue of ingress of the materials composing them is not apparent, nor is any connection with other bodies usually traceable. This feature, together with some other properties of pegmatites, has led to the supposition that many pegmatites arise by the concentration and localization of molten or dissolved matter sweated out of the enclosing rocks. To some proponents of this view, pegmatites represent the first stage in the conversion of solid rocks into fluid magmas by the action of heat and chemical solvents. Magmas thus generated have intrusive power and can later crystallize into igneous-appearing rock masses.

Most geologists, however, regard pegmatites as the end of a lengthy sequence of events rather than the beginning. They consider that as minerals separate from an igneous magma, the water and other volatile constituents such as boron and fluorine become concentrated in the still liquid portion. This highly mobile fluid, from which the pegmatites crystallize, is expelled from the magma chamber into the surrounding rocks. With it go all the elements present in the original magma that have not found a home in the rock-making minerals. It includes such light elements with small atoms as boron, lithium and beryllium, and heavy elements with large atoms such as tungsten, niobium, tantalum, thorium and tin. The atoms of these elements are either too large or too small to enter into the crystal structures of the common minerals of the igneous rock. They thus wind up in the watery, mobile magma that makes its way outward into the country rock and there they enter into the formation of the rare minerals of the pegmatites. The close association in space and time of most pegmatites with major igneous rock masses argues strongly for this latter view. In some areas, however, there are countless small pegmatites but no large igneous masses in the same region. These may indeed be formed by concentration from the surrounding rocks and mark the beginning rather than the end of the long complex process of magma generation and crystallization.

Pegmatites: Shapes and Sizes

Of all rock bodies, pegmatites are probably the least regular and most varied in shape. Some are tubular bodies resembling igneous dikes with fairly regular and clean-cut parallel walls cutting across the structures of the enclosing rocks. The frequency of this habit has led to the practice of calling all pegmatites "pegmatite dikes" although the majority are not dikelike. A very common habit is lenticular, that is, thick in the middle and diminishing in width toward the edge. Lenticular pegmatites are frequently found in layered metamorphic rocks and generally lie parallel to the banding of the enclosing gneiss or schist. A classic example of this type is the great pegmatite forming Collins Hill, northeast of Portland, Connecticut. Here the pegmatite is an elongate lens lying parallel to and wrapped around by the schistosity of the country rock that dips about 45° from the horizontal. Mining operations in Strickland's quarry there have revealed both the upper and lower edges of the lens.

Other pegmatites are shaped like a carrot or an ice cream cone, circular or elliptical in plan and diminishing downward to a point. Others are also circular or elliptical in plan, but remain constant in cross section or even enlarge slightly with increasing depth. These two types are often called "pipes," recalling the kimberlite pipes of the South African diamond fields. However, unlike the diamond pipes, the pegmatite pipes probably do not extend to great depths. Some pegmatites defy any attempt to liken them to any well-known object; they are wholly irregular with flamelike projections and have contacts that in places parallel and in places cut across the layering of the enclosing rock. Others are bedlike, extending between layers of enclosing rock for considerable distance, as is the case of the great pegmatite at Värutrask, Sweden.

In general, pegmatites are small rock bodies and although they vary in size between wide limits, most have a maximum dimension of a few hundred yards. The largest have lengths in excess of two miles whereas the lengths of some are measured only in feet and the width in inches. But in spite of the great variation in size they all are characterized by a coarse-grained texture and frequently a zonal distribution of minerals.

The largest crystals ever recorded have come from pegmatites. A small quarry in the Ural Mountains is reported to have been opened in a single crystal of feldspar, as shown by the continuity of cleavage! A single crystal of phlogopite mica from Ontario has been described as fourteen feet in diameter, thirty-three feet long, and weighing "not less than ninety tons." A well-faced quartz crystal from a Russian pegmatite on display in Moscow, weighs over a ton. The largest reported quartz crystal, weighing over five tons, was found in a Brazilian pegmatite.

Huge crystals in pegmatites are not confined to the common minerals quartz, feldspar and mica. The most famous are the gigantic spodumene "logs" at the Etta Mine in the Black Hills of South Dakota. Here, embedded in a matrix of feldspar, the spodumene lies like giant jackstraws with individual crystals forty to fifty feet long and several feet in cross section. From another mine in the Black Hills a single beryl crystal weighing approximately one hundred tons has been reported. Elsewhere in pegmatites, tourmaline crystals nine feet long, topaz crystals weighing hundreds of pounds and amblygonite crystals several feet across have been found. These are but a few outstanding examples; a complete list of giant crystals from pegmatites would include many of the rarer minerals.

Another characteristic feature of many pegmatites is zoning, or division of the rock body into regions of differing texture and mineral content. A typical pegmatite is wholly enclosed by a narrow wall zone with a relatively fine-grained texture at the contact with the surrounding country rock. The next inner zone, somewhat ambiguously called the "border zone," has a coarser texture, and is generally rich in mica, and often contains black tourmaline. The elongate tourmaline crystals, when oriented at right angles to the contact, sometimes penetrate into the country rock. Grading inward from the border zone are, commonly, one or more concentric, fairly well-defined, though frequently interrupted "intermediate" zones of coarse-grained mixtures of plagioclase and potash feldspar with quartz and mica, in an irregular fabric. Within these "intermediate" zones may be regions composed dominantly of potash feldspar, often in very large blocky crystals and frequently intergrown with quartz as graphic granite. The mineralogy may be complex, with rare minerals such as beryl present in large amount. At the center of the system of enwrapping zones is the "core," frequently of massive quartz, sprinkled with well-formed crystals of feldspar, beryl and possibly containing pockets or cavities lined with crystals. It is the white quartz core of Cerro Blanco that gives that pegmatite its name. The number, regularity and distinctness of the zones varies widely from pegmatite to pegmatite, but the wall zone, border zone, intermediate zone or zones and core are features common to most. It is rather surprising to see pegmatites only a few inches from wall to wall that show these zones clearly, whereas some large pegmatites have but a single well-defined "zone."

Pegmatites: Simple and Complex

It will be convenient to distinguish two kinds of pegmatites: the simple and the complex or "mineralized." The simple pegmatites are vastly in the majority, with a numerical superiority of perhaps one thousand to one. The mineralized pegmatites are usually much larger and what they lack in numbers they make up at least in part in size.

Simple pegmatites are usually small, ranging from a few inches to a few feet in width, and have a mineral composition essentially identical with that of the large igneous rock body with which they are generally associated. Thus, the simple granite pegmatites contain only the rock-making minerals of a granite: plagioclase feldspar, potash feldspar, quartz and a little biotite or muscovite mica. There is a coarse, uneven texture and some zoning is evident at least to the extent of a "core." It is clear that simple pegmatites have little to offer the mineral collector, although a competent explanation of their origin is still a major problem for geologists.

The mineral hunter has a keen interest in the complex pegmatites as the source of many rare minerals and of gem quality specimens of commoner minerals. These are also the pegmatites of economic importance, yielding most of the world's high-grade mica, feldspar and ores of lithium and beryllium. The complex pegmatites are generally fairly large and have well-developed zoning, although the nature and sequence of zones differs from one to another. Vuggy or cavernous textures are common, and it is from these openings that the best crystallized specimens come. Complex pegmatites frequently show a sequence of events in their formation. The common, early-formed minerals may be replaced by unusual minerals, or they may be dissolved, and the resulting

(Above) Labradorite feldspar, ordinarily a dull gray, exhibits a play of colors in this polished slice from Labrador. (Emil Javorsky)

(Below left) These crystals of green microcline feldspar, amazonite, and smoky quartz crystals grew into a cavity in a pegmatite at Pikes Peak, Colorado. (Harry Groom)

(Below right) Rose quartz crystals from Minas Gerais, Brazil, have grown on a base of milky quartz. The cut stone, also from Brazil, appears opalescent. (Emil Javorsky)

cavities may be filled or lined with crystals. If pegmatites are nature's treasure house, these cavities can be called nature's jewel chest, for in them perfectly developed, transparent crystals of a dozen gemmy minerals in many colors have been found.

Many interesting minerals result from the unusual elements in the watery magmas from which pegmatites crystallize. The pegmatites in which they are found are frequently designated by the contained rare element or elements, as, for example, lithium pegmatites, beryllium pegmatites, fluorine pegmatites. In all of these, the common minerals of quartz, feldspar, and mica abound and a pegmatite may be noteworthy for them either because of large or perfect crystals or unusual and gemmy varieties.

Common Pegmatite Minerals

Feldspar is not only the most common and abundant mineral in the rocks but in the pegmatites as well. It is predominantly the potash feldspar, microcline, but in some pegmatites and certain zones of others, albite is an abundant mineral. A snow-white platy variety of albite, cleavelandite, is found only in pegmatites and is usually formed at the expense of earlier feldspar. It is thus present when extensive replacement has taken place in the late stages of pegmatite formation and may be associated with the more exotic minerals.

The humble feldspar in some pegmatites has been touched by nature's capricious wand and transformed into a variety of decorative stones and even into gems. At Itrongay, on Madagascar, potash feldspar (orthoclase) is found as clear pale amber to deep yellow crystals of gem quality. The color apparently results from iron, usually present in feldspars in amounts less than 0.5 per cent, that may constitute nearly three per cent of the Itrongay crystals.

The potash feldspar, microcline, is the most abundant mineral of the pegmatites and may form the gigantic blocky crystals characteristic of many of them. It is usually milk-white or creamy tan but in places the large crystals are colored red by the presence of iron. More rarely the crystals are a beautiful bright green and are called amazonstone or amazonite. Because of its pleasing color, resembling jade, amazonite is used as cut stones in pendants and brooches. The most outstanding localities for this green feldspar are the Ural Mountains, and Pike's Peak, Colorado.

A number of varieties of the plagioclase feldspars are included among the gem and decorative minerals. Labradorite, commonly in small, gray rock-forming grains, is found in a few places in the world in crystals several feet across. In its most notable occurrence, in Labrador, it is also gray, but as the specimen is turned in the light a beautiful series of colors of blue, green, yellow and, less commonly, red plays across the cleavage face. The name peristerite is given to coarsely crystalline albite feldspar showing a similar play of colors; these have been found in pegmatites in Madagascar and in Quebec and Ontario. The gem known as moonstone, with a cloudy appearance and an opalescent play of colors, may be either a variety of albite or orthoclase. Aventurine or sunstone is a feldspar, usually oligoclase, that shows bright red or yellow reflections from spangles of included flakes of iron oxide.

Even though quartz is considered in Chapter 15, it is worthy of mention here for some of its most interesting varieties are the products of pegmatites.

(Above) Rich green tourmaline crystals embedded in milky quartz from Minas Gerais, Brazil.

(Below) Multicolored tourmaline in rose quartz from a pegmatite of Southwest Africa. (Both by Studio Hartmann)

Most spectacular are the huge crystals of water-clear colorless rock crystal that in the state of Minas Gerais, Brazil, have reached the staggering weight of over five tons. However, most pegmatite quartz is not clear but milky-white and as such is the predominant mineral of the core. In some pegmatites the quartz of the core may be a faint pink grading into a deep rose color. This rose quartz is peculiar to pegmatites, and masses weighing many tons have been encountered and quarried for use in monuments. The stone marking the grave of Ralph Waldo Emerson in Concord, Massachusetts, is a great rough block of rose quartz from the Black Hills of South Dakota. Crystals of rose quartz are rare, but small ones of fine quality have come from Newry, Maine; however, the world's most perfect and beautiful are found in cavities in a pegmatite in Brazil.

Feldspar and quartz may be found in pegmatites in a curious intergrowth that was originally called pegmatite but is now known as *graphic granite*. Quartz forms about twenty-five per cent of the intergrowth and is in slender parallel rods within the body of the feldspar crystals. The broken ends of these rods appear on the cleavage surfaces of the feldspar as hook-shaped, L-shaped, or strangely formed figures resembling Arabic or Hebrew characters, hence the name "graphic." It has been suggested that the "tables of stone" that Moses showed the children of Israel were slabs of graphic granite!

The authors once had occasion to delineate the feldspar crystals exposed in the wall of a quarry at Bradbury Mountain, Maine. The task may sound like numbering the sands of the seashore, but it actually was not difficult. In the quarry wall, which measured 120 by thirty feet, there were exposed only 148 blocky crystals with an average cross section of twenty-five square feet. The largest measured eighteen by ten feet in cross section. Part of the interest in these crystals lay in their huge size but of equal interest was the fact that they were all intergrown with quartz as graphic granite. Within one feldspar crystal there may be several clusters of differently oriented quartz rods, but the number of different orientations throughout all the crystals was small. One is, therefore, led to the conclusion that the structure of the feldspar somehow controls the crystallization of the quartz, although precise geometrical "laws" for this beautiful texture have not been formulated.

The mica minerals, muscovite and biotite, so common as small grains in igneous and metamorphic rocks attain their greatest size and crystal perfection in the pegmatites. In addition to these common micas, pegmatites also contain magnesium-rich phlogopite and lithium-rich lepidolite. All the micas are alumino-silicates with a sheetlike structure that confers upon them a single highly perfect cleavage. This property, which permits them to be split into paper-thin sheets, is so typical that the term micaceous cleavage is used to describe a platy cleavage in other minerals. The other physical properties differ slightly from one mica to another, reflecting the differences in chemical composition as shown in the following table.

The Mica Minerals

Mineral	Chemical Composition	Color	Hardness	Specific Gravity
Muscovite (white mica)	$KAl_2(AlSi_3O_{10})(OH)_2$	Colorless	$2-2\frac{1}{2}$	2.76–3.1

Biotite (black mica)	$K(Mg,Fe)_3(AlSi_3O_{10})(OH)_2$	Black	2½–3	2.8–3.2
Phlogopite (brown mica)	$KMg_3(AlSi_3O_{10})(OH)_2$	Brown	2½–3	2.86
Lepidolite	$K_2Li_3Al_3(AlSi_3O_{10})_2(OH)_4$	Purple, pink, gray	2½–4	2.8–3.0

It is difficult for some to picture these minerals, seen usually as thin sheets, as crystals having a large third dimension. However, in well-formed mica, when the whole crystal is seen, it is characteristic to find the greatest dimension at right angles to the cleavage. This was shown in the giant phlogopite crystal from Ontario that measured fourteen feet parallel to the cleavages but thirty-three feet across. Such prismatic crystals usually display a hexagonal or diamond-shaped cross section.

In complex pegmatites the lithium-bearing lepidolite is a late mineral frequently replacing the earlier-formed minerals as a fine, flaky aggregate. In places large plates of colorless muscovite are found with the rim replaced by purple lepidolite.

An interesting phenomenon shown by some, but not all, phlogopite from pegmatites is asterism, resulting from the presence of rutile. During the crystallization of the mica at high temperatures the structure incorporated the element titanium but on cooling could not tolerate it. The titanium was thus rejected by the mica and crystallized in the form of rutile, titanium dioxide, in slender needle-like crystals. The needles were not randomly oriented but, through the constraining influence of the phlogopite crystal structure, were arranged in a hexagonal pattern. When a single light source such as a candle flame is viewed through a thin cleavage flake of this mica, the light diffracted by the oriented rutile needles forms a beautiful six-pointed star.

Minerals of the Complex Pegmatites

The fluid magmas from which pegmatites crystallize have in widely scattered places in the world concentrated large amounts of rare elements, and the pegmatites may be designated by the name of the concentrated rare element or elements such as lithium, beryllium, boron, phosphorus, fluorine tantalum, uranium, tin and cesium. These pegmatites thus contain a great variety of minerals into which these elements are incorporated.

Boron is present in tourmaline; beryllium in beryl and chrysoberyl; fluorine in apatite and topaz; lithium in spodumene, lepidolite, amblygonite, triphylite and petalite; cesium in pollucite and gem beryl. The list of phosphate minerals is long, with 237 described in *Dana's System of Mineralogy* 7th edition, and more than half have been found in pegmatites. Notable localities are the Palermo Mine, Groton, New Hampshire, the pegmatite quarry at Branchville, Connecticut, and Hagendorf, Bavaria. Many of these rare phosphates are formed as alteration products of earlier minerals such as apatite, triphylite and amblygonite, and frequently occur as crystal crusts lining minute cavities in feldspar. The platy uranium phosphates autunite and torbernite occur as coatings on joint planes or sandwiched in between the sheets of mica in a large mica "book." Individual crystals of some species may never exceed a millimeter in length, and some species are represented only by a single specimen from a single locality.

The rare-earth pegmatites, such as the historic occurrence in the quarry near the little town of Ytterby, in Sweden, which gave its name to four chemical elements, contain still other rare minerals such as samarksite, allanite, fergusonite, microlite and gadolinite. It was in a specimen of one of these minerals from Ytterby that the Finnish chemist Gadolin found a new "earth" in 1794, which later yielded the metal to be known as yttrium. Mosander, one of the assistants of the Swedish chemist Berzelius, isolated from Gadolin's impure "yttria" two new earths, which he called "erbia" and "terbia," both names also derived by Ytterby. A Swiss chemist, de Marignac, in 1878, picked a fourth element, which he called "ytterbium," out of the samarskite.

It is in the complex pegmatites that crystals of the gem minerals are most prevalent and on them have been focused the attention of the mineral collector, gemmologist and industrial mineralogist. Although the list could be greatly extended, the most interesting of these minerals are given below.

Minerals of Complex Pegmatites

Mineral	Chemical Composition	Color	Hardness	Specific Gravity
Spodumene	$LiAlSi_2O_6$	White, pink, green	$6\frac{1}{2}-7$	3.15
Petalite	$LiAlSi_4O_{10}$	White	$6-6\frac{1}{2}$	2.4
Eucryptite	$Li(Al, Si)_2O_4$	White	$6\frac{1}{2}$	2.66
Lepidolite	$K_2Li_3Al_3(AlSi_3O_{10})_2(OH)_4$	Purple, pink, gray		2.8–3.0
Amblygonite	$LiAlFPO_4$	White	6	3.0–3.1
Triphylite	$LiFePO_4$	Gray	$4\frac{1}{2}-5$	3.5
Brazilianite	$NaAl_3(PO_4)_2(OH)_4$	Yellow	$5\frac{1}{2}$	2.98
Tourmaline	Complex silicate	Black, various colors	$7-7\frac{1}{2}$	3.0–3.25
Pollucite	$Cs_4Al_4Si_9O_{26} \cdot H_2O$	White	$6\frac{1}{2}$	2.9
Beryl	$Be_3Al_2Si_6O_{18}$	Colorless, green, yellow	$7\frac{1}{2}-8$	2.76
Chrysoberyl	$BeAl_2O_4$	Green, yellow	$8\frac{1}{2}$	3.75
Topaz	$Al_2(SiO_4)(F, OH)_2$	Colorless, yellow, blue	8	3.4–3.6
Columbite-tantalite	$(Fe, Mn)(Nb, Ta)_2O_6$	Black	$6-6\frac{1}{2}$	5.20–7.95

If a mineral is embedded firmly in a matrix, earth movements passing through it may cause it to fracture. Under these conditions a mineral formed originally as a clear, transparent crystal may be criss-crossed with cracks, when uncovered millions of years later, and therefore be worthless as a gem. Thus the best gems are found as free-growing crystals in the cavities and open cracks of pegmatites.

A polished slice of graphic granite (from Bradbury Mountain, Maine) so called because the quartz (dark markings in the cross section) in the intergrowth of feldspar and quartz resembles printed or written characters. (Emil Javorsky)

Tourmaline and Others

Most pegmatites that show the development of any minerals other than the invariable potash feldspar, quartz and mica contain white albite feldspar (cleavelandite) and tourmaline. The chemical composition of tourmaline is given as "complex silicate" in the table. Ruskin in *Ethics of the Dust* concurred with this when he described it as: "A little of everything, there's always flint and clay and magnesia in it; and the black is iron according to its fancy; and there's boracic acid, if you know what that is, and if you don't, I can't tell you today, and it doesn't signify, and there's potash and soda; and on the whole, the chemistry of it is more like a mediaeval doctor's prescription than the making of a respectable mineral." If "lithia" is substituted for "potash" Ruskin's description is not inaccurate. A scientific description of the chemistry of this remarkable mineral would excel Ruskin's in precision but hardly in eloquence.

Tourmaline is usually in elongated crystals with a cross section resembling a spherical triangle and it is coal-black due to the presence of iron. When lithium enters into its structure, tourmaline becomes transparent and variously colored, most commonly green. If rose-pink to red, it is called *rubellite*, if green, *elbaite*, and if blue, *indicolite*. A rare colorless tourmaline is achroite. A most remarkable feature of some gem tourmalines is a color zoning indicating a change of chemical composition during growth of the crystal. "Watermelon" tourmaline is the name appropriately given to a crystal with a red center and green "rind." Elongate crystals may show a color zoning from base to tip, as in one collected by the authors from the famous Gillette quarry near Haddan Neck, Connecticut. It is deep green at the base, shading into lighter green above, then into rose-pink and finally colorless tourmaline at the tip. Individually, cut tourmalines make superb gem stones in rings and pendants, and joined together into a necklace, form a jewel of barbaric splendor.

The discovery of most gem minerals from pegmatites in the United States has been incidental to the mining of feldspar and mica. One of the noteworthy exceptions was the mining of gem tourmaline at Mt. Mica, near Paris, Maine. The discovery of this deposit was made by two boys, Elisha L. Hamlin and Ezekiel Holmes, who had for some time been interested in the minerals that commonly occurred in the fields and ledges near the Hamlin home. One day in the fall of 1820, while hunting for specimens, they were attracted by the gleam of a green crystal caught in the roots of an upturned tree. They secured the mineral, but a heavy snow prevented a further search for minerals until the following spring. By that time, the green crystal had been identified by Benjamin Silliman, professor of chemistry and mineralogy at Yale University, as tourmaline, and with the first thaw the boys returned and secured many fine gems. For more than one hundred years, the locality was worked extensively, and some of the gems from the quarry have found places in the crowns of kings. There is still occasional activity at Mt. Mica, but the great days of gem mining there seem to be ended. Many of the finest specimens of the collection of tourmaline assembled by Augustus Hamlin, son of the co-discoverer, are on view in the Harvard University Museum.

The lithium pegmatites may contain a score of minerals in which lithium is either a major or minor constituent. Of these, spodumene is most frequently

The deeply etched crystal and the two pale gem stones are hiddenite, the rare, green variety of spodumene. The deep green crystals and cut stones are peridot, gem olivine.
(Studio Hartmann)

found, usually as chalky white crystals but occasionally as gem material of rare beauty. At Pala, California, Minas Gerais, Brazil, and Madagascar, gem spodumene of a delicate lilac-rose to amethystine-pink has been found. These transparent crystals of great beauty and value display to a marked degree the property of dichroism; that is, the intensity and quality of the color depends on the direction in which light passes through the crystal. This lovely gem has been given the name kunzite, for G. F. Kunz, a noted American gemmologist. As the variety hiddenite, spodumene has a striking emerald-green color. Hiddenite was discovered as a result of the overturning of a tree on a farm near Stony Point, North Carolina, by W. E. Hidden, for whom the gem variety was later named. The spodumene gems are rare and of exceptionally lovely color, but the pronounced prismatic cleavage makes them difficult to cut and polish, and the relatively low hardness makes them impractical in a ring mount.

Beryl, found in giant crystals firmly embedded in pegmatites, is usually a yellow-green to blue-green color, but the smaller flawless gem crystals are found in several colors to which variety names are given. They characteristically occur in well-formed hexagonal prisms commonly with brilliant crystal faces, but sometimes are etched and corroded by natural solutions, giving the crystals a fluted appearance.

The best-known variety is aquamarine, the transparent blue to sea-green gem. The name heliodor is given to lovely golden-yellow beryl from Southwest Africa, but similar-colored stones from elsewhere are called golden beryl. Morganite is a pale pink to rose-red variety occurring in tabular crystals. It is believed that small amounts of cesium are responsible for the tabular habit and small amounts of manganese produce the pink color. Although some gem beryls from pegmatites are green, they never have the deep green color of emerald. This most precious variety of beryl is considered in Chapter 14.

Although chrysoberyl is found in several types of deposits, some of the finest examples of this gem material have come from pegmatites. The name comes from the Greek *chrysos,* meaning golden, but this is misleading, for chrysoberyl is most commonly a yellowish green and frequently other shades from nearly white to deep emerald-green. Alexandrite is the name given to a variety of chrysoberyl that has the remarkable property of appearing emerald-green by daylight and red by artificial light. This stone was discovered in Takovaya, in the Ural Mountains in 1833 and was named alexandrite after the Czarevitch, later Czar Alexander II.

Some chrysoberyl, because of the presence of parallel needle-like inclusions, shows, when polished as a cabochon, a band of light that moves as the stone is turned. This property led to the popular name of "cat's eye" for such gems, and the term chatoyancy for the effect. "Cat's eye" chrysoberyl is not to be confused with "tiger's eye," a variety of quartz with a golden brown color and a fibrous structure that produces a very prominent moving band of light on a properly polished surface. Cat's eye is actually a more valuable and less common stone than tiger's eye. Chrysoberyl is also called cymophane, from Greek *kyma,* meaning a wave, in allusion to the pale opalescence that forms the cat's-eye effect. Alexandrite, with its dramatic color change—described as "an emerald by day and a ruby by night"—is a valuable gem, but the color change has been successfully imitated in synthetic spinel so the would-be buyer of such stones should beware.

Cut stones and crystals of gem spodumene from Minas Gerais, Brazil. The small crystal of kunzite, the pink variety of spodumene, is sharp and well formed but the larger crystal has been deeply etched by natural solutions.
(Studio Hartmann)

Not a year passes that mineralogists do not discover several hitherto unknown mineral species. But as the science moves forward, discovery shifts from the obvious to the more obscure minerals. Today, it is not unusual for the description of a new mineral to be based on such a small amount of material that practically all of it is consumed in making the chemical analysis. It is thus remarkable when a new mineral is found in large quantity, and even more remarkable when it is found in large crystals of gem quality. But such was the discovery of the mineral brazilianite.

In 1944, F. H. Pough, an American mineralogist in Brazil, was shown by a mineral dealer a large yellow-green crystal presumed to be chrysoberyl. Dr. Pough immediately noted that neither the hardness nor the crystal symmetry agreed with those of chrysoberyl. Eventually, the optical properties, x-ray diffraction pattern and chemical analysis showed it to be unlike any known species. It came from a pegmatite in Minas Gerais, Brazil, and the name brazilianite was given it. Since its discovery, many additional specimens have been found in Brazil and cut into gem stones. It has also been found elsewhere in pegmatites, notably in New Hampshire, but in small crystals and not of gem quality. Although brazilianite makes an attractive gem, it has relatively low hardness (5½) and luster and therefore cannot compete with golden beryl in either durability or brilliance.

Precious Topaz

Although topaz is found with the tin ores of Cornwall and Nigeria, and in cavities in rhyolite lava in Utah and Mexico, its most noted occurrences as a gem are in pegmatites. Pure topaz is colorless, but small amounts of impurities color it yellow, blue, green, brown and red. In the gem trade, topaz is considered to be of a wine-yellow color and goes under the name "precious topaz" to distinguish it from the more abundant and less expensive citrine quartz. Most of the precious topaz has come from Minas Gerais, Brazil, where it is found in deeply weathered pegmatites and as water-worn stream pebbles. Beautiful, well-crystallized blue topaz has come from pegmatites in the Ural Mountains.

Topaz takes its place with other minerals which are found in large crystals in pegmatites. During World War II, in a search for quartz crystal in Brazil, a dozen large, colorless, transparent crystals of an unknown mineral were uncovered. After the war they were identified as topaz and sold to museums in the United States and Europe. The largest of these crystals weighs 596 pounds and may be seen in the American Museum of Natural History in New York City.

In 1929, the Department of Mineralogy of Harvard University leased a pegmatite property at Topsham, Maine, in the hope of finding good crystals in cavities. After weeks of work a small blast at the bottom of the quarry broke into a cavity containing a blue mineral, topaz. At the time of formation, in the distant past, the topaz had crystallized as several large, pale blue crystals that would have made fine exhibit material. However, since then, solutions had attacked the topaz, leaving only a skeletal boxwork to indicate the former size and quality of the crystals. The mineralogists had arrived perhaps a million years too late!

Industrial Minerals of Pegmatites

In spite of their small size, unpredictable character and varied mineral content, pegmatites have real economic importance. They constitute the world's chief source of high-grade feldspar, electrical-grade mica, ores of the metals beryllium, lithium, niobium and tantalum and some of the piezoelectric quartz.

Potash feldspar and to some extent the lower-temperature melting albite feldspar are used in great quantities by the ceramic industries. Although the total tonnages of feldspar shipped annually are large, each mining operation is, by the nature of pegmatites, with their low tonnages and limited continuity, a small venture. For this reason much feldspar mining has been an intermittent family business carried on in conjunction with farming, fishing, lumbering and other activities. But in recent years, with the advance of technology of mineral separation, large mining operations are being carried out on pegmatite bodies. Now the whole pegmatite is quarried, finely crushed, and the several minerals separated from each other. Marketable feldspar, quartz and muscovite mica result from this operation. Although the individual small operator is still active, his days appear to be numbered.

Feldspar is used in the manufacture of whitewares and dinnerwares, in the glass industry and as a mild abrasive cleanser. Freedom from iron is a most important consideration for these applications, and thus black tourmaline and biotite mica are removed magnetically. The feldspar is then ground in ball mills using flint cobbles from the beaches of Normandy, in order to avoid contamination by iron during grinding.

Beryl and Cryolite

In the past, only the gem varieties of beryl have been of interest. As the gem *smaragdus*, it was among the stones most valued by the ancients, and it is also found in medieval works of alchemy as a specific remedy for belching! Only since the 1920's, with advancing metallurgical technology, has it taken on additional significance as an ore of beryllium metal. For this purpose it makes little difference whether the crystals are large or small, flawless or cracked, colorless or green. It is the tonnage that counts. Although beryl is found in places in gigantic crystals, it is usually scattered irregularly through the pegmatite in crystals not more than a few inches in length. There are no great beryl mines and the world's production of the mineral comes from many small operations involving much sorting and hand picking of material. As is true with many other pegmatite minerals, Brazil is the chief producer of beryl. Further, low labor costs make possible recovery of the mineral from relatively lean deposits.

Beryllium is a light metal (specific gravity, 1.85) but has almost the same structural properties as aluminum (specific gravity, 2.70). It thus finds many uses in aircraft, missiles and space vehicles where weight is an important factor. Because of its transparency to radiation, it is used in nuclear reactors and x-ray equipment. Further, it is an important alloy metal in beryllium copper and confers high strength and resistance to corrosion on bronzes and aluminum alloys. The present world's production of beryl is small, being less than ten thousand tons, of which the metal beryllium itself makes up only about

360 tons. The price is therefore high, over sixty dollars a pound for the metal. The uses would be quickly extended if the supply were greater and the price lower. Research is being carried out actively in the hope of recovering beryllium from other minerals such as phenacite (Be_2SiO_4), bertrandite ($Be_4Si_2O_7[OH]_2$) and helvite ($Mn_4Be_3Si_3O_{12}S$).

Fluorine is not an uncommon element in pegmatites but is usually present in small amounts in such minerals as apatite and topaz and more rarely as fluorite. However, in the great deposit at Ivigtut, at the head of Arksukfjord in southwestern Greenland, fluorine is an abundant element. Here several fluorine minerals are present, but the most important is cryolite, a sodium aluminum fluoride. For a pegmatite, this deposit contains an unusual assemblage of minerals: siderite, an iron carbonate; galena, the chief ore of lead; and sphalerite, the chief ore of zinc, all embedded in a matrix of cryolite.

Although the deposit had long been known, it was not until the middle of the nineteenth century that attempts were made to mine it. The galena contained small amounts of silver and the deposit was opened as a lead-silver mine, but the operation was unsuccessful.

Then in the 1880's the electrolytic process for the separation of aluminum from bauxite was independently discovered in France by Heroult, and in the United States by Hall. The solvent in which the bauxite was dissolved, so that electrolysis could be conducted, was molten cryolite. The great deposit at Ivigtut then came into its own, and active mining has continued to the present day. Mining is difficult because of the rigors of the harsh climate, and shipment of ore is restricted to a brief period each year during which Arksukfjord is free enough of ice to permit the entry of ships. But the cryolite is easily extracted and the deposit enormous, a great pipe widening downward. So climate notwithstanding, mining has gone on. Today much of the cryolite used in the electrolysis of aluminum is synthetic, prepared from the mineral fluorite. Small amounts of cryolite and other complex fluorides have been found in pegmatites in Colorado, in the Pikes Peak district, and in the Ural Mountains; elsewhere in the world it is found only as a curiosity.

Mica

With the exception of some finely ground mica, the world's supply of this important industrial mineral comes from pegmatites. India is the chief producer, followed by Brazil. This does not mean that deposits in these countries are the richest or most extensive but that the prevailing low labor costs permit careful preparation of the material for market. For commercial purposes sheet mica must be a good electrical insulator and withstand high temperatures. The two mica minerals that meet these qualifications are muscovite and phlogopite. Lepidolite melts at a relatively low temperature and biotite with its high iron content is an electrical conductor.

Before the manufacture of glass, one of the chief uses of muscovite was as window panes. "Books" of mica could be split into clear, thin sheets and then trimmed to the appropriate size. One can still see windows in old palaces in India glazed with muscovite that let in the light and keep out the cold winds today as effectively as they did when they were constructed. This white mica was mined in Muscovy (the early name for Russia) for this purpose and was

(Above) Orthoclase feldspar and brazilianite are among the rare yellow gem minerals, the yellow orthoclase of gem quality coming only from Madagascar and the gem brazilianite only from Brazil.

(Below) Blue-green aquamarine, a gem beryl from Minas Gerais, Brazil; and golden beryl from Easthampton, Connecticut. (Both by Emil Javorsky)

called Muscovy glass; hence the name of the mineral. After the advent of glass for house windows, the chief use for muscovite was as windows in furnaces and heating stoves where it was necessary for the window to be both transparent and heat resistant. At this time it was usually referred to by the German term, *Eisenglas,* corrupted in English to "isinglass." Although today sheet mica is still used for furnace windows, the demand is small. The greatest consumption is by the electrical industry in punched forms for radio tubes, for winding electrical resistance elements, and for high-grade capacitors. Sheet mica for such purposes is becoming increasingly rare and sells for high prices—from a few dollars to seventy-five dollars a pound. Because of the enlarging demand and the dwindling supply of high-grade mica, synthesis of sheet mica is being actively carried out and may bring new life to the industry.

The mica unfit for sheet purposes as well as the scrap from punchings is finely ground and used as roofing material, paint and wallpaper.

Lithium Minerals

Lithium is an important element in many pegmatites, but in a few places in the world concentration of this element has reached gigantic proportions. Notable localities are at Värutrask near Boliden, Sweden; Bikita, Southern Rhodesia; and Karabib, Southwest Africa. At these places hundreds of thousands of tons of the lithium aluminum silicates, spodumene, petalite, eucryptite and the lithia mica, lepidolite, have been concentrated. Although not containing lithium, the cesium mineral, pollucite, elsewhere rare, may be present in thousands of tons. It is, however, in those pegmatites with somewhat lesser concentration of lithium that the gemmy lithium-bearing minerals are found.

Lithium is one of the important "space age" metals for which new applications are being found daily. It is light, highly reactive, and has the simplest nuclear structure among the metals. Many lithium compounds are used in industry, but the most important is lithium stearate. As such it is used in greases that lubricate equally well at high and low temperatures—from $-60°$ to $320°$ F.

Niobium (called columbium in newspaper metal market columns) and tantalum are almost wholly won from the minerals columbite and tantalite, which occur exclusively in pegmatites. These are minerals of high specific gravity and may collect in placer and eluvial deposits. The chief world producers are Nigeria and the Congo where they are recovered from decomposed granite and associated pegmatites. Tantalum is noted for its high melting point, mechanical strength and chemical inertness and is used for laboratory ware, electron tubes and, like gold, for use in contact with living tissue as cranial plates and other surgical aids. Columbium is used in alloys that must withstand high temperatures, as in jet engines and gas turbines.

(Above left) Twinned crystal of chrysoberyl from Minas Gerais, Brazil.
(Katherine H. Jensen)

(Above right) Topaz occurs in well-formed crystals of various colors: (top left) blue crystals with albite feldspar from Siberia; (top right) a colorless crystal from Southwest Africa; precious amber-colored topaz, the smaller crystals (bottom left) from the Thomas Mountains, Utah, and the larger crystal (bottom right) from Minas Gerais, Brazil.

(Below) Cryolite, an icy-looking mineral, enclosing black galena and green-brown siderite. Cryolite is found abundantly only in Ivigtut, Greenland. (Both by Emil Javorsky)

6 A Traprock Suite: Zeolites and Related Minerals

Stretching across most of eastern Oregon and the southeastern quarter of Washington is a vast upland or plateau, built up some thirty million years ago by outpourings of volcanic lava on a scale seldom equaled. This is the Columbia Plateau, more than 200,000 square miles of volcanic tableland split by vast gorges through which flow the great Columbia River and its eastern tributary, the Snake.

At Hell's Canyon, the Snake River has cut downward for over a mile vertically giving a vertical exposure of flow upon flow of basaltic lava in the grim walls of a gorge deeper than the Grand Canyon of the Colorado.

In Oregon, in the heart of this volcanic tableland, a hundred miles to the west of Hell's Canyon, is Ritter Hot Springs. Here, at the bottom of a small crater-like pit, hot water steams and bubbles odorously. On either hand are jagged walls of black basalt. Cavities in the basalt are numerous, some bubble-like, others mere hairline fractures. Black widow spiders inhabit some of the cavities and have lined them with the untidy, pearly floss of their webs. In other cavities, tufts of snow-white crystals shimmer wondrously against the background of jagged black basalt. The tufts, as much as two inches in diameter, are made up of thousands of crystals, each so delicate that it trembles in the breath of the viewer. These natural productions seldom survive efforts to remove them for the vibration of a hammer blow will reduce the slender crystals to glittering ruin. However, study of the contents of the bubble-like cavities with a magnifying glass reveals that with the slender fibers occur other minerals as crystalline crusts and isolated crystals. To name them all with confidence may require a laboratory examination using a polarizing microscope and x-ray diffraction equipment. But with these instruments we can identify the slender fibers of the tufts as mesolite, which is a zeolite, and the associated

(Above) Blocky apophyllite crystals with zeolites in a cavity of a Brazilian lava flow. Smaller cavities are completely filled with zeolites. (Studio Hartmann)

(Below) Minerals of the traprock suite: (right) pink apophyllite crystals and white to gray sheaf-like bundles of stilbite from the Ghats Mountains, India; (far right) green prehnite and black babingtonite crystals coating a vein in traprock from Westfield, Massachusetts. (Emil Javorsky)

minerals as three other zeolites: chabazite, analcime and stilbite and calcite.

Such an association of minerals is not confined to Ritter Hot Springs but may be seen wherever dark volcanic rocks abound. What makes Ritter Hot Springs particularly interesting is that zeolites seem to be forming as a result of present activity in the springs. The composition and temperature of these waters give us a precious clue to the geological environment prevailing during the deposition of other zeolites associated with volcanic activity of the distant past. Zeolites have been synthesized in the laboratory at temperatures between 100°–500 °C (212°–932 °F), thus supporting the testimony of geological environment in fixing the conditions of origin. The warm alkaline solutions circulating through the rock, deposit zeolites and related minerals in cavities of the rock from which the necessary chemical ingredients were leached. These beautifully crystallized minerals were probably precipitated from waters so dilute that their only evidence of dissolved mineral matter would have been a rather "hard" taste.

Geology of the Traprock Suite

The home of the zeolites is in the basalts, fine-grained volcanic rocks, which since the beginning of geologic time have been periodically erupted onto the earth's surface. These dark, nondescript lavas are called traprocks and in cavities within them are found the minerals of the traprock suite. When lava is extruded, and confining pressure is lost, the water and volatile constituents that helped keep the lava molten during its journey to the earth's surface escape very rapidly. These gases bubble out of the lava like carbon dioxide from a freshly opened bottle of carbonated water. Some of these bubbles are trapped by the quickly solidifying lava producing a rock containing many spherical or oval cavities somewhat like Swiss cheese. Such a rock is said to be vesicular, and the gas cavities are called vesicles.

The lava shrinks on cooling, producing fractures called columnar joints that divide the rock into prismatic masses or columns. The forces of weathering pry these columns apart and tumble them down to form picturesque outcrops resembling ruins, flights of steps or piles of cordwood and castles and have generated some of the most evocative names on the land: the Giant's Causeway, the Devil's Postpile and the Palisades are but a few.

Heated waters rising from below and surface waters percolating downward along fractures and openings dissolve sodium, calcium, potassium, aluminum and silica from the traprock. The same cavities and fractures may then provide a place for the deposition of new minerals from the solutions, the minerals of the traprock suite.

Traditionally, mineralogists have thought of zeolites as rather rare minerals occurring in delicate crystals in traprock cavities. But recent work has revealed them to be present in huge deposits although in far less attractive specimens. At numerous desert lakes in California and Nevada, volcanic tuffs originally composed of shards of volcanic glass have been found to be almost wholly altered to a hard, chalky-looking rock consisting mainly of zeolites. Similar zeolitization of volcanic ash beds has been reported from Olduvai Gorge in Tanzania. The most abundant zeolites resulting from this process appear to be phillipsite and erionite. The active agent in such mass zeolitization is

(Above) The Ajanta caves, the oldest dating from 200 B.C., were hollowed out by a Buddhist order in the Deccan traps, a series of horizontal basaltic lava flows that cover half a million square miles. (Edward S. Ross)

(Below left) Amethyst. Quartz crystals (full size) from Guerrero, Mexico, grade from colorless at the base into the rich purple amethystine variety at the top. (Reo N. Pickens, Jr.)

(Below right) A nearly spherical "water" agate found filling a cavity in a lava flow in Brazil. (Studio Hartmann)

simply the warm, highly saline subsurface water of the playa lakes, or the scanty alkaline ground water of the hot, dry African valleys. The time required for this major alteration is surprisingly short from a geologic point of view, estimated at a few hundred to a few thousand years.

Apparently when volcanic glass crystallizes in the presence of alkaline sodium and calcium-bearing ground water, zeolites are generally the product. Consequently, many species hitherto considered rare must now be regarded as abundant and geologically significant.

Properties of Zeolites

Zeolites are a well-defined group of closely related minerals, and the mineralogist has little trouble in spotting them in their characteristic association. A recent compilation (1963) lists twenty-seven zeolites that give different x-ray diffraction patterns and may be viewed as distinct species. To this number must be added numerous synthetic zeolites having no natural equivalents.

All zeolites are white or colorless when free from impurities, but are occasionally colored flesh, pink, tan or gray by the inclusion of oxides of iron or other foreign mineral matter. Their luster is feeble, lacking any "fire" or brilliance, and gives the delicate crystals an ethereal quality contrasting with the harsher glitter of their mineral neighbors. The specific gravity is notably low; the average of twenty-seven species is only 2.2, much lower than that of quartz or feldspar. All the zeolites can be scratched by the steel of a good knife ($H = 5.5$) yet are harder than the fingernail ($H = 2.5$). Most display one or more cleavages, with a pearly luster on many of the cleavage surfaces.

The most striking property of zeolites, although rather difficult to observe, is the dehydration, or water loss. When most water-bearing compounds are heated, no change takes place until a definite temperature is reached, then

Stilbite crystals from Nova Scotia form characteristic sheaflike aggregates. (Wilbert Draisin)

water is lost rapidly. When a zeolite is heated, water is given off at once, well below 212° F (the boiling point of water) and continues to be lost progressively as the temperature is raised without significant change in the crystal structure. The dehydrated crystals can then reabsorb most of all of the lost water when cooled. If heated hot enough, however, the zeolite will fuse, usually with the bubbling and curling that led Baron Cronstedt, a Swedish mineralogist, to give to the group its name from the Greek *zeo* meaning I boil. Because of their avid thirst for water, dehydrated zeolites make excellent drying agents. They have, moreover, the virtue that after becoming fully water saturated, they may be restored to serviceability by simple heating. This property of reversible hydration-dehydration is so typical of zeolites that any mineral containing water which may be lost and regained without structure change is said to contain "zeolitic water."

Although some zeolites occur in vast quantities as alteration of volcanic glass, they are massive and have little appeal to the mineral collector. But in their traprock home they are usually found in beautifully formed crystals and can be divided for easy recognition into four groups based on their characteristic crystal habits. Some zeolites are far more common than others, but where the common zeolites occur, rare ones may be present also.

Natural Zeolites

Group 1. Fibrous: natrolite, scolecite, mesolite, thomsonite, gonnardite, edingtonite.
Group 2. Bladed: heulandite, stilbite, epistilbite, dadiardite, brewsterite.
Group 3. Equant: analcime, chabazite, phillipsite, harmotome, gmelinite, faujasite, levyne.
Group 4. Miscellaneous: erionite, laumonite, mordenite, gismondine, ashcroftine.

Characteristics of Zeolites

Natrolite is the most widespread of the fibrous zeolites, and may be taken as typical of the group. It is an aluminum silicate high in sodium, and thus derives its name from the Greek *nitron* meaning soda. It forms slender prismatic crystals with square cross section, often needle- or hairlike, and commonly joined in radiating clusters and divergent groups. Although usually small, single crystals nearly three feet long and four inches wide have been reported from eastern Quebec. In most of the localities where it is found, it is one of the last minerals to crystallize and hence is found as tufts and groups implanted on the earlier minerals, or as fibrous masses filling the space between earlier-formed minerals.

Scolecite, mesolite, gonnardite and edingtonite, although differing chemically, closely resemble natrolite in their crystal structure and hence in their habit and properties. In addition to the usual radiating prismatic crystals, thomsonite forms massive compact fillings of gas cavities in lava which may weather out to form spherical or almond-shaped pebbles with radial structure and concentric color banding. Water-worn pebbles of this kind, weathered from the traprock of the Keweenaw Peninsula, Michigan, and polished to an agate-like appearance in nature's tumbling mill of the beach, are found on the shores of Lake Superior. The colors are white, reddish brown and green.

Heulandite and stilbite are distinct species, each with its characteristic crystal habit and structure, but with similar chemical composition. Both are widespread and typical of the bladed zeolites. The crystals generally assume a more or less elongate lozenge or tablet shape with a highly perfect cleavage parallel to the flattened surface of the tablets. Heulandite and stilbite are most often observed in the classical setting of the traprock cavities, but heulandite is also reported as a constituent of massive bedded deposits of altered volcanic ash in New Zealand, Australia and California. Stilbite is not uncommon as a late-formed mineral in crevices and on joint faces in metamorphic and igneous rocks and is recorded in association with epidote, prehnite and feldspar in fractures in gneiss and mica schist.

Heulandite is commonly found as lozenge or coffin-shaped single crystals flattened parallel to the prominent cleavage, which shows a distinct pearly luster. Single crystals of stilbite flattened parallel to the pearly cleavage and looking like little Roman short-swords are sometimes seen, but aggregates are much more common. Most frequently these take the form of nearly parallel or slightly diverging bundles resembling a tied sheaf of wheat or a head of cauliflower. The brown color of some stilbite adds to the wheat sheaf illusion, and the old name for stilbite, desmine, is from the Greek *desme*, a bundle, in allusion to this habit.

Analcime and chabazite have in common their generally stout, blocky, equidimensional habit, and although as distinct mineral species they differ in chemical composition and structure, may both be placed in the arbitrary but convenient grouping of the equant zeolites. Analcime crystallizes in crystals with twenty-four sides of almost spherical appearance, called trapezohedrons, which may be water-clear, as in the basalts of County Antrim, Ireland, or white, as in the traprock quarries of New Jersey. Sometimes analcime occurs alone implanted directly on the black surface of basalt, looking like frozen droplets of water. More commonly it is associated with other members of the traprock suite such as calcite and datolite, looking like big white snowballs with tufts of natrolite and crystals of heulandite and stilbite clustered around them. Analcime plays another role in nature, as a primary constituent of certain rare igneous rocks, of which it may make up twenty per cent.

Chabazite forms rhombohedral crystals, sometimes as large as an inch across, which differ so little in appearance from cubes that when chabazite was first described as a mineral species in the eighteenth century, it was thought to be cubic. The cleavage parallel to the rhombohedron faces is not easy or prominent, chabazite differing from other zeolites in this respect. The crystals from the Bay of Fundy district in Nova Scotia are yellow or flesh-pink to red in color rather than the usual white, and were called acadialite at one time. The name chabazite seems to have originated in a copyists' error. Ancient authors refer to *chalazios* or *chalazias,* and Agricola mentions chalazias in 1546 as a stone resembling hail in shape. The use of a name suggesting hail is a pleasant conceit for a mineral whose water-clear rhombohedra sometimes resemble hailstones. But sometime between 1546 and 1788 the "*l*" became transformed into a "b."

Non-Zeolite Minerals of the Traprock Suite

The zeolites are rarely found alone. Not only are different members of the family usually found together, but other non-zeolite minerals are their characteristic and frequent associates. The entire suite of minerals so frequently occurs together associated with large bodies of basaltic and diabasic igneous rock (traprock) that the association has been called the traprock suite.

Non-zeolite members found almost exclusively in cavities of traprock are: prchnite, datolite, apophyllite and pectolite. Amethyst, calcite, native copper, native silver, epidote, babingtonite and chlorite although characteristic of the suite are frequently found in other associations as well.

Prehnite is a green mineral with a divergent-bladed habit, generally forming compact crusts with an undulating or kidney-shaped surface. Datolite characteristically forms equant, glassy, colorless to light green crystals having many faces. There is no cleavage, and the glassy appearance, complex crystals, moderate hardness and slightly greenish color make recognition easy. Some of the world's finest datolite has come from the giant trap quarries just east of Westfield, Massachusetts. Datolite is also commonly associated with the famous Lake Superior copper ores. Here it is found as fine-grained aggregates having the appearance of cauliflower, which when broken yield a fracture surface resembling white, unglazed porcelain. Some of these massive datolite aggregates contain finely divided native copper scattered uniformly through them. The delicate red color imparted by the copper makes these specimens some of the most treasured of the Copper Country minerals.

Apophyllite, with its dainty pseudocubic crystals, its prominent cleavage with pearly luster, its low specific gravity and white color is certainly a close relative of the zeolites and a worthy member of the traprock suite.

Pectolite has a fibrous habit, much like natrolite, which it somewhat resembles, and generally forms radial globular masses or crusts with rounded surfaces. Individual fibers are sharp-pointed and splintery and have a habit of getting into the skin of inquisitive mineralogists.

The violet variety of quartz, amethyst, is a common associate of zeolites and frequently is found in crystals of high quality that are cut as gem stones. Calcite is possibly commonest of all and may occupy cavities in traprock unaccompanied by any other mineral. Copper and silver are occasionally found as fancifully twisted wires and thin sheets but usually are minor members of the mineral association. In places, however, they have been found so abundantly as to be mined for ores, as the copper on the Keweenaw Peninsula, and the silver at Kongsberg, Norway. Hematite, epidote, chlorite and many other minerals, such as the rare babingtonite, are found as occasional constituents of the traprock suite. Although babingtonite is elsewhere a rare mineral, it is found at seven places in Massachusetts, the outstanding locality being Westfield. Here it is found as glittering black crystals perched on a background of green prehnite and partially embedded in later calcite.

Where Zeolites Are Found

Wherever large quantities of dark igneous rocks are found, one may look for the traprock suite. Some obvious places are in northern Ireland, where the columnar basalts come marching down to the sea to form the Giant's Causeway, offshore islands, and sea caves like Fingal's Cave that so greatly impressed the young Mendelssohn. In County Antrim, there are hundreds of localities where zeolites and related minerals are found in the joints and cavities of traprock.

On the night of November 15, 1963, a narrow black ridge of volcanic ash broke the surface of the sea near the barren Westman Islands off the southwest coast of Iceland, and the island of Surtsey was born. Continuing eruptions of the volcano Surtur, named for a destructive Icelandic giant, built the island during the next year and a half, not only with ash and bomb falls but with streams of lava flowing into the sea. In this way, if activity continues, an island like the Hawaiian islands will be created of shield form and constructed of layers of hard lava. Where the waves of the Atlantic break on the still-hot solidified lavas of Surtsey, mingling cold sea water and volcanic emanations, zeolites are probably already in the process of formation, and the new island may one day be a noted locality for these flowers of the mineral kingdom, as are the older islands of the Icelandic group.

On the American side of the Atlantic the waves break on traprock headlands in southern Nova Scotia in the Bay of Fundy area, where sea cliffs reveal the zeolite-bearing lavas at Cape Blomidon, Peter's Point, Isle Haute and Wasson's Bluff, popular areas for picnicking, rock fishing and mineral hunting. One must be careful not to be caught by the prodigious tides of the Bay of Fundy on a sea-cut bench with an unscalable wave-cut cliff above!

But more famous and most accessible of traprock areas are those of the so-called Newark troughs of the eastern seabord. From Nova Scotia to South Carolina, parallel to the Atlantic coast, there stretches a chain of basins formed by downfaulting some two hundred million years ago, in Triassic time. As they sank, these basins were filled by red sediments swept into them by torrential streams from the neighboring highlands. Periodically, volcanic eruptions poured fresh lava out on the surface of the accumulating sediments, and injected dikes and sills of basalt and diabase into the thick series of sediments already accumulated. Further faulting tilted the entire mass of sediments with included lavas. Erosion has since removed part of the red sedimentary rocks and exposed the tilted edges of the resistant dark basalt flows, sills, and dikes, which today stand above the surrounding countryside as prominent ridges and hills. In nearly every such ridge, extending from the Watchung Mountains of New Jersey to the Holyoke Range of the Connecticut Valley in Massachusetts, is today a traprock quarry. In each of these quarries, operated for the excellent road metal the trap makes, there is the potential for interesting finds of zeolites or other minerals of the traprock suite.

Descending as in giant steps from the central plateau of India, the traveler crosses a series of basaltic lava flows, the Deccan traps. From cavities within these rocks have come the large and handsome specimens of zeolites that grace most of the mineral museums of the world.

Zeolite Behavior

A very important aspect of zeolite behavior, deriving from the curious structural chemistry of the group, is their capacity for base exchange. The "bases" or

(Above) Spherical aggregates of pectolite associated with zeolites occur in a traprock cavity at West Paterson, New Jersey.

(Above right) Large pseudocubic crystals of chabazite and white calcite line a cavity in black basalt at West Paterson, New Jersey. (Both by Wilbert Draisin)

metal ions other than aluminum and silicon are held in the interstices and pores of the open aluminosilicate framework by weak electrical bonds. Consequently, metal ions of similar size may replace each other in these sites without disturbance of the framework. Thus, if a sodium zeolite such as natrolite is soaked in a solution containing calcium ions, the calcium and sodium will change places. The calcium enters into the crystal structure while the sodium goes into the solution. This process called base exchange or ion exchange is used to soften water. "Hardness" in water results chiefly from dissolved calcium ions. Thus, hard waters are generally found in regions where the underlying rocks are limestone or sandstone with a lime cement. The dissolved calcium ion reacts with household soap to form insoluble soaps, or "scum," which have no cleansing power and leave a ring in the sink or bathtub. Hence, to use soap effectively in hard-water areas, it is necessary to "soften" the water; that is, to get rid of the calcium ions. This may be done by passing the water through a tank containing fine-grained sodium zeolite, essentially the natural zeolite, natrolite. The calcium enters the natrolite structure, and the sodium from the zeolite goes into the water. The water becomes "soft" because sodium does not form insoluble soaps or scum. When the zeolite becomes saturated with calcium, the tank is replaced with a fresh one. The calcium-saturated tank is not discarded but is rejuvenated by passing through a strong brine of sodium chloride, that is ordinary salt. The sodium replaces the calcium in the crystal sites, making the tank ready for service again. With such periodic rejuvenation, water softeners can operate for a long time without rebuilding.

Base exchange is not confined to sodium and calcium. During the fixing operation in the processing of photographic film, the "hypo" (fixing bath) eventually loses its effectiveness by becoming saturated with silver. If a finely ground zeolite is immersed in the "hypo," the silver ions will substitute for other ions in the zeolite structure and can be recovered later.

Cesium and rubidium, strontium, barium and thallium all enter zeolite structures from solution and, like the calcium of the water-softener tanks, may be displaced at a later time by brine treatment. Consequently, zeolites are undergoing study as "packaging," storage and shipment media for radioisotopes of exchangeable ions, as well as a possible agent in the fixing of dangerous radioisotopes in the disposal of atomic wastes.

So characteristic of zeolites is this base-exchange capability, that a field test using this property has been suggested to distinguish fine-grained zeolites from other minerals that resemble them.

When a zeolite has been thoroughly dehydrated by heating, its great thirst is generally slaked by absorption of moisture from the air. If, however, no water is available, zeolites can absorb an astonishing number of substances. These include the oxygen and nitrogen of the air, ammonia, iodine vapor, bromine vapor, mercury vapor and a variety of organic materials both gaseous and liquid. Furthermore, this process is selective. For example, if a mixture of iso-octane and higher hydrocarbons derived from the distillation of petroleum is passed through a bed of properly chosen zeolite at the correct temperature, the iso-octane will be selectively absorbed and the higher hydrocarbons rejected. The iso-octane can then be expelled from the saturated zeolite by heating, resulting in a high concentration of octane by means of a simple and relatively inexpensive procedure. Because of this behavior, the name "molecular sieves" was applied to the zeolites as early as the 1920's.

Molecular sieving involves the use of a dehydrated zeolite in which the pores or passages between the large cavities in the aluminosilicate framework are of such a size as to admit one component of a mixture while excluding others. This can be done in part by the choice of the zeolite. The pores in some zeolites such as faujasite are very large, whereas in others, such as natrolite, they are small. The size of the pores can also be controlled to some extent by adjusting the temperature of the sieving operation; the higher the temperature, the larger the effective pore diameter. When calcium substitutes for sodium, one doubly charged calcium ion can do the same job of maintaining electrical neutrality as two singly charged sodium ions. Since these atoms partially block the pore entrances, and obviously one ion gets in the way less than two, the substitution of calcium for sodium increases the effective aperture. By adjusting these factors, the molecular sieving process can be made finely selective.

Pursuing possible industrial application of molecular sieves, mineralogists and chemists have synthesized numerous zeolites, some of them the equivalents of natural minerals, but most of them unknown in nature. These synthetic zeolites find a ready market. As drying agents they are extremely effective, reducing water content of gases to levels as low as four parts per million; as molecular sieves they find many applications such as enriching gasoline in octane; and as carriers for such costly catalysts as platinum they increase the effectiveness by even distribution while retaining the catalyst.

7 Crystals in Sedimentary Rocks

Limestone, composed mainly of calcite, is a common sedimentary rock underlying large areas on every continent. It is also the most soluble of the sedimentary rocks. Rain water with carbon dioxide in it will dissolve limestone at the earth's surface at the rate of about one foot in two hundred years. On a geologic time scale, this is extremely rapid. Large amounts of calcium carbonate are thus constantly carried to the sea to be deposited once more as limestone and once again to begin the cycle.

In limestone regions it is not only the surface of the rock that is dissolved. As the rain water works its way downward, it enlarges the cracks in the rocks through which it moves, forming large channelways for underground drainage. The extensive systems of caves and caverns found in many parts of the world have been formed in this way.

If water with its dissolved calcium carbonate emerges as a spring from a hillside, some of the calcium carbonate forms a porous deposit called travertine or calcareous tufa, made up mostly of calcite. The calcareous material encrusting the objects over which the water flows is frequently the petrifying material of fossil plant and animal remains. Most travertine is a gray friable or crumbly mass possibly colored by small amounts of impurities. The town of Yellow Springs, Ohio, gets its name from a travertine deposit colored a yellowish red by a small quantity of iron. Calcium carbonate separated from the warm water emerging from the ground at Mammoth Hot Springs in Yellowstone National Park has formed the famous vari-colored terraces there. At Carlsbad (Karlovy Vary), the celebrated spa in Bohemia, Czechoslovakia, similar deposits of *Sprudelstein* have formed from the numerous hot springs. Although most travertine is soft and crumbling, some is cohesive and hard enough to be cut and polished for use in floors and wainscoting in public buildings. Such decorative material comes mostly from Tivoli near Rome, Italy. It is the principal building stone of Rome and is used for decorative purposes throughout the world.

After channels and cavities have appeared in a limestone rock, a decrease in rainfall may cause precipitation, that is, the formation of deposits rather than further solution. Only small amounts of water may enter the cavern and drip slowly from the roof. Evaporation will then cause the calcium carbonate in the water to build stalactites downward from the roof and stalagmites upward from the floor. If they meet, a column is formed. Cave deposits may have the appearance of hanging draperies, festoons, or other fanciful shapes. Water seeping on the floor of a cavern may build up a layered translucent deposit resembling a frozen cascading waterfall.

Although much less common than calcite, the mineral aragonite, another form of calcium carbonate, may be precipitated as cave deposits. In some Mexican caverns alternating layers of calcite and aragonite form an attractive decorative material. Fashioned into small objects it is sold under the name of Mexican onyx. Similar material used by the early Egyptians was called "Oriental alabaster."

Long after the formation of a cave, solutions rising from depths may encrust the floor, roof and walls with well-formed crystals containing elements completely foreign to the surrounding limestone. The best exhibit specimens of many minerals have been removed from such crystal-coated caverns.

Calcite

An ancient series of adjoining caverns exists in the region surrounding the meeting point of Missouri, Kansas and Oklahoma. This area, known as the Tri-State district, has long been a major producer of lead and zinc. The ore minerals galena and sphalerite were deposited in rock cavities only a few hundred feet beneath the surface. The district has not only furnished superb specimens of these minerals but equally fine specimens of many others. The mineral marcasite, which is iron sulfide (FeS_2), is found in exceptionally fine crystal groups in this district. It is found in stalactites covered with crystal faces; or it may occur in aggregates called "cockscomb marcasite." In the same association, pinkish crystal crusts of dolomite with a pearl-like luster called "pearl spar" coat crystals of minerals formed at an earlier time. Sprinkled at random on all the crusts are tiny, well-formed, brass-yellow crystals of chalcopyrite. Similar deposits in limestone of such crystallized minerals are known in many parts of the world, most strikingly at Bou Beker, Morocco on the Algerian border, where mining of lead-zinc ore has revealed the same associations.

Calcite, a common associate of ore minerals, is found in many mining districts. Some of the most famous localities for beautiful crystals are: Cumberland, Durham, Lancashire and Cornwall in England; the Harz Mountains and the silver mines of Freiberg and Schneeberg in Germany; and Příbram in Czechoslovakia. But in no place is calcite found more abundantly in well-formed crystals than in the American Tri-State district. The region is especially noted for clear to amber-colored calcite crystals of many shapes and sizes; the long dimension of some crystals is as much as six feet. Since the mineral is of

The principal mineral of cave deposits, calcite assumes fantastic shapes, as in the stalactites and stalagmites in Mammoth Cave, Kentucky. (Ray Scott: National Park Concessions)

Calcite crystals plucked from a cavern in Arizona, each crystal showing perfect development of several crystal forms. (Floyd R. Getsinger)

little interest to miners, many caves lined with calcite have remained intact. No experience can be more thrilling to a mineralogist than to enter such a cavern and see the light of his torch reflected from a myriad of crystal faces.

Calcite is unique in the diversity of its crystal forms. When chemically pure, it is clear and colorless and called Iceland spar from its occurrence in cavities in lava flows near Helgustadir on Eskefiord, Iceland. It is more commonly milky white but may be tinted red, green, blue or yellow. Interesting crystals of calcite enclosing grains of sand are found in the Bad Lands of South Dakota and at Fontainebleau in France. These "sandstone crystals" are formed by calcite crystallizing in sand. A single crystal continuity may extend for several inches, incorporating as much as sixty per cent quartz sand.

Clear calcite shows better than any other common mineral the property of double refraction, that is, the breaking of light into two rays, each of which produces a separate image. In 1828 William Nicol, taking advantage of this double refraction, constructed a prism from calcite that eliminated one of the rays, permitting the other to emerge as plane polarized light. Ordinary light vibrates in all directions at right angles to its path; plane polarized light also vibrates at right angles to the path, but in a single direction. As an integral part of the polarizing microscope, this "nicol prism" gave a great impetus to

(Right above) Calcite crystals encrusted with an iron oxide have been partially coated with a transparent amber-colored calcite.

(Below) Pink dolomite sprinkled with small brass-yellow crystals of chalcopyrite, a copper ore, from a lead-zinc mine near Joplin, Missouri. (Both by Studio Hartmann)

the study of minerals and an understanding of the optical properties of crystals.

One of the most noticeable features of calcite is an excellent rhombohedral cleavage that permits it to break into rhomboidal fragments. The first speculation regarding the internal structure of crystals was brought about by a study of this cleavage. In 1782, René Just Haüy, a professor at the University of Paris, was horrified when a friend accidently shattered a prized calcite crystal by dropping it on the floor. However, the accident had its rewarding side, for Haüy noted that both large and small fragments had the same shape, and suggested that the cleavage rhombohedron is the primitive form out of which all calcite crystals are built. By analogy he suggested that other crystals are also built of primitive forms of varying shapes, a premise that with modifications is held today.

Aragonite

Although aragonite shares with calcite the chemical formula $CaCO_3$, it is completely different in terms of crystallography. It forms slender tapering crystals called "church steeple" crystals by English miners. More commonly three crystals of aragonite grow together forming so-called pseudohexagonal twins. Such twin crystals are found in the province of Aragon, Spain, from which the mineral gets its name. Similar and even more perfectly formed crystals are found in the sulfur deposits at Girgenti, Sicily.

Aragonite is much less common than calcite, forming under a narrower range of pressures and temperatures and is usually found in deposits near the surface. In some iron ores as in the Styrian deposits in Austria it occurs in white, elongated stems resembling coral and is called *flos ferri* (flower of iron). The nacre or mother-of-pearl of many shells, and pearl itself, is aragonite.

Gypsum Crystals

Hydrous calcium sulfate, that is, gypsum, usually occurs, as described in Chapter 8, in large deposits as a precipitate from sea water; but in some places it is found as free-growing crystals in rock cavities. In the state of Chihuahua, Mexico, there are many metallic ore deposits in which limestone caverns are closely associated with the ore; and in some areas the presence of caves is considered a positive indication that ore lies below. The formation of the caves is believed due in part to the normal action of surface waters and in part to the dissolving action of circulating waters containing sulfuric acid, generated by oxidation of sulfide ore minerals. Although some caverns contain calcite crystals similar to those of the American Tri-State district, others are lined with crystals of gypsum deposited by the sulfate waters.

The most remarkable display of gypsum crystals was found in a cavern of the Maravilla Mine at Naica, Chihuahua in northern Mexico. Dr. William Foshag of the United States National Museum has described the floor of the cave as littered with blocks of limestone sloughed off from the roof, coated with calcite and sprinkled with long crystals of gypsum. In the grottos between the limestone boulders, he reported, there are usually clear selenite crystals like figures in niches. The crystals from the roof and upper walls resemble icicles but those in the lower parts of the cave are clear, colorless and well-formed.

(Above left) Pointed crystals of aragonite from Cumberland, England, are called "church steeple" crystals by local miners. (Katherine H. Jensen)

(Above right) Although aragonite, from Aragon, Spain, is orthorhombic, three of its crystals may grow together to form a nearly hexagonal prism. (Arthur Twomey)

(Below) Aggregates of delicate aragonite crystals stained blue by copper have grown on a background of tyrolite, a green copper carbonate, from Schwaz, Austrian Tyrol. (Harry Groom)

In a lower chamber the crystals have grown to enormous size, many of them four and five feet long. Further on, a narrow passage, lined with bladelike crystals, forms a "veritable corridor of swords." Beyond this, in the largest chamber of the cave, the floor rises at an angle of about 30° and is completely banked with selenite blades one to three feet long. At the high point of the floor there arises a radiating group of gray crystals over four feet high, tipped with white and glistening under the lamps. The large radiating group to which Dr. Foshag refers was removed from the cave crystal by crystal and reassembled at the Harvard University Mineralogical Museum.

In 1892, Dr. J. E. Talmage of Salt Lake City identified as gypsum some thin, transparent mineral sheets sent him from the Wasatch Plateau, Wayne County, Utah by a sheepherder who thought that they might have a commercial value. The gypsum was the clear variety, selenite, and of such high quality that Dr. Talmage decided to investigate. When he arrived at the location he found a mound thirty-five feet long, ten feet wide and twenty feet high that had been left on a hillside after erosion had removed the sand and clay around it. One end of the mound had been broken off, revealing a cavern extending the length of the mound. Clear crystals from one to four feet long projected inward from the sides, the largest extending like great beams entirely across the cavern. Dr. Talmage promptly acquired the property to preserve the crystals as mineral specimens. More than thirty tons of crystals, each weighing from ten to more than one hundred pounds, were removed, and one can see specimens from this unique occurrence in every sizeable mineralogical museum in the world.

Fluorite and Luminescence

Fluorite, calcium fluoride, is found in many geological environments. It is deposited in veins with metallic ore minerals, especially those of lead and silver; it may be an accessory mineral in igneous rocks and pegmatites; it is found with barite, gypsum, celestite and dolomite, or it may occur in veins with no other mineral. But some of the largest and best-formed crystals of fluorite are found extending into cavities formed by solution in limestone rocks.

Fluorite has many interesting properties and uses. It crystallizes in cubes that are sometimes intergrown to form twinned crystals. Many of the finest crystals have come from the lead mines of Cumberland and Derbyshire, England. The miners there early learned that the most desirable specimens were not only well-crystallized but exhibited unusual crystal faces. To accommodate the purchasers, the miners frequently supplemented the common faces by others carefully filed and polished.

Fluorite has perfect octahedral cleavage; that is, there are four directions of easy breaking. Thus, with patience, one can cleave a perfect eight-sided octahedron from a fluorite cube. When pure, fluorite is transparent and colorless, but it is rare in this state. A variety of impurities and structural imperfection give rise to different colors. Crystals may be green, yellow or purple or, more rarely, blue, pink and brown.

The impurities, coupled with imperfections in the crystal structure, cause fluorite to be remarkably luminescent. Luminescence, the property of some minerals to emit visible light other than by incandescence, can be created in

These terraces at Mammoth Hot Springs, Yellowstone National Park, Wyoming, are layers of calcium carbonate precipitated from the cascading hot water. (Tozier Collection, Harvard University)

When calcite crystallizes from water permeating loose sand, the sand grains may be incorporated into the growing crystals, as in these specimens from the Badlands of South Dakota. (Wilbert Draisin)

(Facing page, above left) A rosette of tabular crystals of barite, from Texas, enclosing sand grains and colored by iron oxide. Such "barite roses," resulting from the weathering of limestones, are common. (Benjamin M. Shaub)

(Above right) Mauve-colored crystals of fluorite from Cave-in-Rock, Illinois, are partially encrusted with colorless crystals of calcite.

(Below) A crust of lustrous crystals of cerussite, from Morocco, is partially obscured by hemispherical pinkish aggregates of barite. (Both by Studio Hartmann)

different ways. Thus, triboluminescence is caused by crushing or scratching, thermoluminescence by heating, and fluorescence by exposure to radiation of short wave length. All of these types of luminescence are generally exhibited in fluorite but it is impossible to predict which specimens will luminesce. The only way to find out is to try, but the trial should be made in the dark, since sometimes the light is extremely feeble. In a mine with the light extinguished, triboluminescence will cause a glowing trail to remain for many seconds after a rock face of fluorite has been scratched by a prospector's pick. If a small piece of fluorite heated slowly in a test tube begins to glow before the mineral reaches the temperature of red heat, it is thermoluminescent. The color of the light emitted differs from specimen to specimen and may be yellow, green, blue or violet.

Fluorescence, which derives its name from fluorite and is the most familiar type of luminescence, is produced by exposing the mineral to ultra-violet light. Many minerals have this property and well-chosen specimens produce a spectacular exhibit. Even ordinarily dull minerals suddenly come to life and give off brilliant colors (see Chapter 12).

Fluorite occurs in thousands of areas, in some only as scattered grains, in others as deposits of thousands of tons. The few places where it has grown out into an open space furnish the most spectacular specimens. These are usually

well-faced crystals but an interesting type, found at Derbyshire, England, is a fibrous and banded blue variety called "Blue John." This material was known to the Romans, who used it for carving cups, bowls and vases, and ever since then it has been used for ornamental objects. Because of its fibrous nature, Blue John is not as fragile as single crystals and thus easier to work. But single crystals have been carved; in fact, the Chinese, using the green variety resembling jade in color, have made exquisite statuettes and images out of it. Some extraordinary carvings in fluorite, produced by the American Indians, have been excavated in the Angle Mounds in southwestern Indiana.

The most beautifully crystallized fluorite has been found in England, where it occurs in perfectly formed, blue and violet cubic crystals, sometimes encrusted with small quartz crystals. The best-known of the many areas in which fluorite has been mined in the United States are at Rosiclare in southern Illinois and adjacent parts of Kentucky. At famous Cave-in-Rock, a few miles up the Ohio River from Rosiclare, where the fluorite lines and partially fills limestone caverns, it occurs in a variety of colors and has been deposited in well-formed cubes, some as large as a foot on one edge.

As the most common mineral containing fluorine, fluorite (CaF_2) is an important raw material for the chemical industry. It is used commercially, under the name fluospar, in the preparation of hydrofluoric acid essential in the manufacture of synthetic cryolite used as an electrolyte in recovery of aluminum metal. It is also the source of fluorine for organic fluorides used as refrigerants and inert fluorocarbon resins used for pipe and tank linings. One such product used for lining cooking utensils goes under the trade name Teflon. The largest amount of fluorite is used by the steel industry as a flux in open-hearth furnaces

(Left) Tabular barite tinted with iron oxide on a background of cerussite crystals from Morocco. (Studio Hartmann)

A nearly spherical aggregate of crested barite from the Harz Mountains, Germany, is composed of tabular crystals. (Cornelius S. Hurlbut, Jr.)

to produce a fluid slag which cleanses the steel of phosphorus and sulfur. A small but important use is as lenses in special optical equipment. For this purpose, fluorite must be clear, colorless and quite free of impurities. But synthetically grown crystals are now increasingly used to satisfy the demand for high-quality optical material.

Barium and Strontium Minerals

Barite (barium sulfate) has an unusually high specific gravity (4.5) for a non-metallic mineral; in fact, it gets its name from the Greek word meaning *heavy*. Well-formed crystals are frequently found with ore minerals in metalliferous veins and as fillings in cavities in limestones and other sedimentary rocks. The most highly prized specimens come from Cumberland and Westmoreland, England. When taken from the rock cavities, the crystals are usually brown but exposure to sunlight for a few months turns them a light blue. In places an intergrowth of tabular crystals forms rosettes that are known as barite roses.

Barite is the principal source of barium for the chemical industry. But its chief use, amounting to over one million tons a year is as a weighting agent for the mud used in drilling oil wells. The inert barite with its high specific gravity greatly increases the weight of the mud, preventing oil and gas from being blown from the hole.

Witherite (barium carbonate) is much less common than barite but occurs in similar associations. It is most commonly crystallized in pseudohexagonal twins, the finest specimens of these crystals coming from the now exhausted lead mines of Cumberland, England.

Celestite (strontium sulfate) resembles barite so closely that it is impossible to distinguish them by inspection alone. They have similar crystal structures, with the position of faces, angles between them, and most physical properties

(Above) On the Wasatch Plateau in Wayne County, Utah, in 1892, a sheepherder came upon a large knob (right) leading into a cavern lined with gigantic clear crystals of gypsum. The largest group to be removed, shown here (left) in a photograph by J. E. Talmadge, weighed 750 pounds and had individual crystals of over five feet in length.

nearly identical. Celestite gets its name from the fact that the first crystals described were a celestial blue, but it is commonly white or colorless.

Celestite is usually found throughout limestones, sandstones and shales or in thin layers between them. The mineral mined from these bedded deposits is the principal ore of strontium. More rare, but of greater interest to the mineralogist, are well-formed crystals of celestite found lining cavities in sedimentary rocks. The world's most remarkable occurrence of celestite is on the island of Put-in-Bay in Lake Erie. A farmer discovered it in 1897 while drilling a well for water. At a depth of seventeen feet, his drilling bit suddenly dropped into an underground cavern. The driller assumed he had encountered a limestone cave similar to others on the island. To recover his drilling equipment he dug into the cave, and to his astonishment found himself in a grotto twenty-five feet long, fifteen feet wide, and twelve feet high completely lined with pale-blue crystals, some of them eighteen inches long. The crystals were celestite of a size and perfection not known anywhere else in the world. Soon after "Crystal Cave" was discovered, the owners decided to preserve it as a natural wonder. Only a small number of crystals have been removed and few museums have specimens.

Aside from celestite, the only strontium mineral of commercial importance is the carbonate known as strontianite. Its occurrence is much more restricted and, as it commonly contains calcite and galena as impurities, its industrial uses are limited. Its crystals are usually needle-like and although generally white, may be yellow, green or gray. Strontianite occurs in veins in limestone along with barite, celestite and sulfide ore minerals.

Much of the strontium recovered from celestite and strontianite is converted to compounds used in signaling flares. The element supplies "the rocket's red glare" and the red light of fireworks and highway warning flares. As strontium hydroxide it is important in the refining of beet sugar. Because of impurities in molasses, only about half of the sugar in it crystallizes; if the molasses syrup is mixed with strontium hydroxide, the remaining sugar can be recovered from the strontium saccharate which results.

8 Minerals from the Sea

The sea, mother of all life, ultimate destination of all sediments, and highway of mankind, is also the source of great quantities of minerals important to man. Sea water is a complex solution and, as shown by the table below, contains about 3.5 per cent dissolved salts, chiefly sodium chloride. Depending on the conditions at the time of deposition the various elements may combine to form compounds other than those listed. For example, potassium sulfate may well form instead of potassium chloride.

Composition of Sea Water

Compound		Compound	
Water	96.24%	Calcium sulfate	0.14%
Sodium chloride	2.94%	Sodium bromide	0.06%
Magnesium chloride	0.32%	Potassium chloride	0.05%
Magnesium sulfate	0.15%	Calcium carbonate	0.01%

Iron oxide—less than 0.001%

For centuries man has been extracting some of these compounds by the evaporation of sea water, but nature has been doing it on a large scale throughout most of geological time, forming sedimentary deposits known as evaporites. From time to time, "geological accidents" have caused a body of sea water to be cut off from the open ocean, as by the development of a bay mouth bar, or the building of a delta across the entrance to a large bay. As evaporation slowly reduces the volume of water in such a body, additional water enters periodically from the sea, but the dissolved salts become more concentrated. This process is most effective in an arid region where there is little rainfall and

Irregularly intergrown crystals of gypsum with inclusions of iron oxide encrust the walls and floor of a cave at Los Lamentos, Mexico. (Floyd R. Getsinger)

evaporation is very rapid. A classic example of such a concentration is the Kara Bogaz Gulf on the eastern side of the Caspian Sea. The Gulf is almost isolated from the Caspian Sea by a sandbar with a shallow inlet, so that water can enter the Gulf but nothing can pass out. Here gypsum and sodium sulfate are being deposited today and if the inlet becomes entirely closed other salts will shortly begin to crystallize. Environments of this kind in the geologic past have produced most of the large deposits of salt, gypsum and sylvite being mined today.

If the volume of such a cut-off body of sea water is reduced by evaporation to about half; that is, if the salinity is doubled, calcium carbonate and iron oxide precipitate. Since there is little of either of these in average sea water, it is obvious that most calcium carbonate (limestone) and iron ore deposits are not formed in this way. Actually, most limestone is formed from the shells, tests or hard parts of organisms, or by the activity of organisms in secreting and concentrating calcium carbonate. Sedimentary iron ores, likewise, owe their origin chiefly to activities of iron-secreting or iron-concentrating organisms.

When the original volume of trapped sea water has been reduced by evaporation to about twenty per cent, gypsum is deposited. As the salinity increases, anhydrite may succeed gypsum, or if the temperature is sufficiently high, anhydrite may deposit from the beginning without prior gypsum deposition. These two minerals are both calcium sulfate, but gypsum contains water, whereas anhydrite, as the name implies, lacks water. That their conditions for deposition are not very different is indicated by thick deposits in which they are sometimes intimately interstratified. After formation and burial beneath other sedimentary rocks, gypsum can change to anhydrite by losing its combined water; and anhydrite can change to gypsum under the attack of the weather, or at shallow depth in the presence of water. When anhydrite changes to gypsum, a thirty per cent volume increase takes place, sometimes doming up the rocks above the altering anhydrite, and sometimes forcing the beds of anhydrite into intricate folds.

When continued evaporation reduces the enclosed body of sea water to ten per cent or less of its original volume, sodium chloride, the mineral halite or rock salt, begins to deposit. So great is the amount of this material in sea water that a relatively small body of water like the Gulf of Kara Bogaz, with a volume of about forty-eight cubic miles today, could deposit 2700 million tons of salt. If evaporation continued and almost all the water disappeared, the bittern salts, potassium chloride, magnesium chloride and magnesium sulfate, together with a great variety of double salts, may be precipitated.

Laboratory experiments on sea water seem to establish the sequence of deposition of oceanic salts given above: first, calcium carbonate and iron oxide, next gypsum and anhydrite, then sodium chloride, and finally salts of potassium and magnesium. Occurrences of these minerals indicate that the situation in nature is often much more complex than that which prevails in the Gulf of Kara Bogaz or in the experimenter's evaporating dishes. We frequently find extensive deposits of one mineral unaccompanied by others as well as deposits in which the predicted order is reversed.

Further, simple evaporation of an enclosed, shrinking body of sea water could never have furnished the great thickness of many salt deposits and bedded sequences of gypsum, anhydrite and salt, such as those in the great Carlsbad

(Above) The dunes of White Sands National Monument, New Mexico, are composed of grains of gypsum. (Emil Muench)

(Below) Ordinary salt, halite, usually white or colorless, is colored deep-blue by radioactive emanations. (Studio Hartmann)

These salt crystals (from California) are called "hopper-shaped" because material added to the edges left a depression in each cube face. (Emil Javorsky)

deposits in New Mexico or at Stassfurt, Germany, which exceed one thousand feet. Evaporation of over fourteen hundred feet of sea water is necessary to yield a single foot of anhydrite or gypsum. Over twenty-five miles of sea water would be required to yield one hundred feet of gypsum or anhydrite. This means that the evaporating basin must have been repeatedly replenished during a very long period of evaporation under fairly constant climatic conditions of aridity and rapid evaporation. The great salt, gypsum, and anhydrite deposits of the world were formed during two periods of geologic time, the Silurian, about 400,000,000 years ago, and the Permian, about 200,000,000 years ago. The conclusion seems inescapable that much of the earth must have had a climate that was arid or semiarid for long periods of time during both of these ancient eras. The salt deposits of New York, Ohio and Michigan are Silurian, those of Texas, Kansas and New Mexico and Stassfurt, Germany, are Permian. The name Permian for this division of geologic time is taken from Perm, U.S.S.R., where there are also extensive salt deposits.

The Salt of the Earth

The mineral halite, common salt, is a physiological necessity for man or beast, and hence has always had a more urgent relation to man's affairs than those minerals sought for their beauty, magical powers, or even their usefulness. Any human group so unfortunate as to lack a supply of salt had to barter for it or fight for it. Consequently the records of human commerce indicate salt as one of the commodities transported over the earliest caravan routes. It was probably second only to flint among the minerals to enter the human trade pattern. Frequently the salt traded was impure, containing large amounts of clay and other minerals. But so valuable was it that it was commonly stolen in transit by a method not immediately detectable: a bag of the mixture could be soaked in water and much of the salt dissolved out of it. When it reached its destination, such "salt," as the Bible puts it, "had lost its savor."

In Egypt, as in many ancient monarchies, the salt trade was a royal monopoly and taxes on it provided substantial royal income. In Roman times, the soldier, if his performance was satisfactory, was given his ration of salt, otherwise it was witheld. We thus have the expression today "A man is not worth his salt." Later, instead of issuing salt, the Roman soldier was given a "salary" with which to buy it. So necessary is salt for man's existence that its unbalance in distribution was probably one of the prime causes of unrest and civil upheaval in the ancient world. The infamous *gabelle* or salt tax is considered to have been a leading cause of the French Revolution.

Fortunately for world peace, salt is abundant and widely distributed, and is produced in more than seventy-five countries. The production in the United States is the largest, constituting more than twenty-five per cent of the world supply, with the states of Michigan, Texas, Louisiana, New York and Ohio the chief producers. Other major producing countries are: England, with most salt coming from Cheshire; the Soviet Union, where it is produced in several widely scattered areas; and Germany, with the most productive areas in Anhalt and in northern Saxony.

As a sedimentary deposit, rock salt beds underlie thousands if not hundreds of thousands of square miles of the earth's surface, and in some places are as much as two thousand feet thick. Some of these come to the surface, and the upper parts are dissolved by circulating ground waters resulting in salt springs. Others are buried beneath many thousands of feet of overlying sedimentary rocks. If the beds are not deep beneath the surface, the common method of extraction is by "brining"; that is, pumping water through a drill hole to the salt-bearing layers and pumping the resulting brine to the surface, a method first used by the Chinese, about A.D. 400. This method has the drawback that to recover the salt the brine must be evaporated, thus introducing the expense of heating.

Although much salt was mined or quarried from earliest times, a large percentage has been obtained since Mycenean-Minoan times by the purposeful evaporation of sea water in salt pans, enclosed ponds to which sea water is admitted and in which it is evaporated by the heat of the sun until salt is precipitated. The salt is then harvested with rakes to separate it from the bittern, or residual liquid rich in potassium and magnesium. About thirty per cent of the salt produced today is recovered from salt pans.

If the salt is not buried too deeply, it may be recovered by sinking shafts and mining the solid mineral underground. In spite of such expressions as "back to the salt mines," implying disagreeable working conditions for those banished to salt mines, a salt mine is rather a pleasant place to work. It is of course dry, usually airy and of an even temperature. It is a thrilling experience to go into the large quiet chambers sometimes two hundred feet high from which salt has been removed and see the clean white walls sparkle in the miner's lamp as the light is reflected from the many cleavage faces. Salt mines too numerous to mention are in operation today but perhaps the most famous is at Wieliczka, Poland (formerly in Austria), which has been worked for centuries. Here unknown sculptors have hewn from the soft massive rock-salt, chapels, galleries and monuments, and have carved statues that can never see the light of day lest they be destroyed in a humid environment. When, in the closing phases of the war, the Nazi art thieves of Hitler's regime wished to hide the plunder from Europe's art galleries, the abandoned salt mines of the Salzkammergut region in Austria and northern Italy were chosen because of their constant temperature and humidity. No art works stored in this way were damaged as a result of their exile.

One hundred miles west of New Orleans on the coast of Louisiana, the islands known as Jefferson, Avery and Weeks lie in a straight line. These are not islands in the normal sense, but ground several score feet above sea level connected by marshy land to the mainland. On Avery Island a salt spring had long been known, and in 1862 this prompted the drilling of a hole in the center of the island to determine the depth of the salt bed below the surface. Salt was encountered at one hundred feet but although subsequent holes penetrated to a depth of five thousand feet, the bottom of the "bed" has never been reached. A hole bored at the edge of the island showed only a shallow deposit at that point. The fact that the whole salt deposit was of great vertical depth yet small in horizontal extent baffled the geologists. Jefferson Island and Weeks Island were found also to be the surface expression of similar salt plugs. But they proved not to be unique. Since that time a systematic search for them has located several hundred *salt domes* along the Gulf coast of Louisiana, Texas and Mexico as well as far out in the Gulf itself. This search was actually undertaken to locate petroleum trapped along the flanks of the salt domes. It is now generally believed that the salt itself was precipitated in horizontal beds from sea water in Permian time and then buried to an unknown depth (greater than seventeen thousand feet) by overlying sedimentary rocks. Under the tremendous weight of the overyling rock the salt moved as a plastic material and punched its way upward, arching the rock through which it passed. On reaching the surface the salt is dissolved and carried away by surface waters, leaving behind the insoluble minerals, mostly anhydrite, as a cap rock. In Iran, in a region of complete aridity, salt that has punched its way to the surface is not dissolved but forms salt glaciers that move downslope as ice does in mountains in colder climates.

To purify the salt mined from salt domes it must be dissolved in water and the impurities, the insoluble minerals, removed. Most of the insoluble residues consist of anhydrite but the other common associates of salt, calcite, dolomite and magnesite, are also present. Less abundant are crystals of quartz, pyrite, goethite, hematite, sulfur, celestite, barite, feldspar and hauerite (manganese disulfide). Occasionally the rare minerals boracite and danburite are found, and in one instance two new minerals, hilgardite and parahilgardite, were discovered.

Most halite occurs in an interlocking aggregate in which the individual grains show no crystal faces. A well-formed crystal must be free to grow without interference; thus occasionally in a cavity in rock salt a clear transparent cubic crystal is found. If the crystallization is rapid a crystal of curious habit is

Water from the Black Sea at Pomorie, Bulgaria, is run into pans where solar evaporation makes the brine continually more salty. After the first salts, calcium carbonate and gypsum, are precipitated, the brine is drawn off into other pans, where the halite crystallizes as salt and is scraped into piles. (J. Allan Cash)

formed, as is well illustrated at Great Salt Lake, Utah. Here, an extensive freshwater lake of glacial times has been evaporating for thousands of years in an enclosed basin in which inflow from neighboring mountains does not equal the rate of evaporation. Consequently, the salinity of the Great Salt Lake today is about thirty-five per cent, or ten times that of the ocean. When the wind whips the surface of this briny sea, the spray evaporates in mid-air to form minute crystals of salt, or evaporates immediately after contact with a solid body to form an adhering crystal. On piles at the amusement park, called Saltair, crystals develop as a result of spray accumulation. In these crystals the cube faces are replaced by stepped depressions that look like angular funnels or hoppers; hence the name "hopper crystals." These skeletal crystals grow so rapidly that only the points that grow fastest, the edges and corners of the cube, fill out; there was insufficient time and material for the more slowly growing faces of the cube to be completed. Snow crystals are also skeletal, and remain incomplete for the same reason.

Most halite is colorless or white, but is sometimes pink, red or brown from included iron oxide impurities, or gray because of included clayey material and dirt. Some chemically pure natural halite, however, is a striking deep-blue,

These transparent gypsum crystals (from Ellsworth, Ohio) were formed in unconsolidated mud apparently without any point of attachment for the growing crystal. (Cornelius S. Hurlbut, Jr.)

called "Prussian blue" because of its resemblance to the iron cyanide of that name. The color is usually not uniform and occurs in patches in otherwise clear crystals of halite from Stassfurt, Germany, and many other localities. This color was for a long time a mystery because of the absence of any chemical or mechanical impurity. The answer to the problem came with the discovery that irradiation with x-rays, subjection to an intense electrical field, bombardment with various kinds of energetic radiation and even heating in sodium vapor could produce the color artificially in pure halite. The color is apparently a result of disordered arrangement of the atoms making up the crystal; an explanation similar to that suggested for the rather similar but more varied colors of fluorite.

Salt is a major raw material of the chemical industry which, in the United States, consumes nearly seventy per cent of the production. The chlorine and sodium extracted from it enter into countless products, many of which are encountered daily. As an essential constituent, chlorine is used in bleaching agents, water purifiers and many insecticides; sodium is used in sodium carbonate and washing soda; sodium bicarbonate, in baking soda; and sodium hydroxide in household lye. Salt is used in large quantities to melt ice on the highways, and in food processing by the canning, baking, fishing, dairy and meat-packing industries.

The human consumption of salt, originally its prime or only use, is today small compared with the great industrial uses. Home use of salt for table and cooking amounts to about fourteen pounds per individual a year, or approximately 21,000,000 tons for a world population of three billion. This is about twenty per cent of a world production of over 100,000,000 tons. However, in industrial countries the percentage for domestic use is much less, amounting to only three to four per cent of the total.

Gypsum: Its Uses

The hydrous calcium sulfate, gypsum, is among the softest of minerals and can readily be scratched by the fingernail. It can thus be worked with tools of stone, metal or bone as easily as can wood or clay. On the streets of Pisa today, for instance, it is not uncommon to see a sidewalk vendor carving with a pocketknife little leaning towers of Pisa, which he offers for sale within a few minutes. The gypsum he is using came, in all probability, from the neighborhood of Volterra, the "lordly Volaterrae" of ancient Etruscan days. Statues, vases, urns and ornaments of the same gypsum may be found, worked no doubt by similar simple methods, in the Etruscan tombs that abound in that ancient land. The pits from which the Etruscan gypsum came, and still comes, may be seen dotting the countryside. This fine-grained, sometimes translucent gypsum with a white to pinkish color is called alabaster. It was as popular with the ancient Egyptians as with the Etruscans, and many of the canopic jars, or funerary vessels used to contain the viscera of the mummified dead man, are made of this material. It should be noted that much of what is called "alabaster" in Egyptian collections is really fine-grained calcite.

Principal Minerals Precipitated From Sea Water

Mineral	Chemical Composition	Hardness	Specific Gravity	Crystal System	Cleavage
Halite	NaCl	2½	2.16	Isometric	Cubic
Gypsum	$CaSO_4 \cdot 2H_2O$	2	2.32	Monoclinic	1 excellent 2 poor 3 directions
Anhydrite	$CaSO_4$	3–3½	2.95	Orthorhombic	
Sylvite	KCl	2	1.99	Isometric	Cubic

Gypsum may grow from the walls of caverns in curved fibrous aggregates, as in this specimen, resembling a ram's horn, from Brewster County, Texas. (Harry Groom)

Alabaster can be colored artificially, and in Mexico many of the "jade" carvings sold to tourists are in fact alabaster dyed green.

Gypsum as bedded sedimentary deposits is as widespread as rock salt and is mined in vast quantities in almost every country. In the Paris basin of France there are extensive beds of gypsum that were exploited at an early time, and the labyrinthine catacombs beneath the city of Paris are the abandoned mines. Most gypsum mined today is converted to plaster of Paris, a name that comes from this old source. The process of making plaster is ancient and was known to the Egyptians five thousand years ago. It consists of "burning" the gypsum, that is, heating it gently to a temperature slightly above that of boiling water to drive off three quarters of the contained water. Thus gypsum, $CaSO_4 \cdot 2H_2O$, goes to $CaSO_4 \cdot \frac{1}{2}H_2O$. When powdered plaster of Paris so prepared is mixed with water, it regains the lost $1\frac{1}{2}$ H_2O and "sets" as a firm recrystallized material. Its uses are many but most of it is consumed in the building trade as wallboard, lath, and sheathing board as well as plaster applied to walls and ceilings. Other important applications are in surgical casts, dental plasters, and molds in the ceramic industry.

Although of less importance than material "burned" for plaster, "raw" or unfired gypsum is used extensively in the manufacture of portland cement, as an agent to retard setting, and in agriculture as a soil conditioner. These and

Each gypsum crystal in this group was carefully removed from a cavern at Naica, Mexico, and reassembled in the Harvard University Museum. (PSSC Physics, D.C. Heath & Co.)

many other uses of both "raw" and "burned" product make gypsum mining and processing a major industry with an annual world production of nearly forty million tons.

Shallow quarries in the neighborhood of Fleurus, a suburb of Oran, Algeria, expose beds of gypsum in which individual crystals are of giant size. The brilliant African sunshine is reflected in blazing splendor from mirror-like cleavage faces several feet in length and as much as a foot wide. These crystals have been formed by recrystallization of sedimentary gypsum through the agency of circulating ground water, and similar large crystals are known from other localities. This coarsely crystalline variety, generally colorless and transparent and displaying the characteristic perfect cleavage of gypsum, is called selenite, after Selene, goddess of the moon. In some shales and related clayey rocks, gypsum crystals of a high degree of perfection occur, bounded on all sides by crystal faces. These crystals probably have grown in the solid body of the clayey rock by the interaction of dilute solutions of sulfuric acid with calcium carbonate present in the shale. Such crystals are to be seen in every mineral collection and generally have the form of a diamond-shaped lozenge with beveled edges. Sometimes two such lozenges are intergrown to form a twin crystal having the shape of an arrowhead, called a "swallowtail twin" from the fancied resemblance of the notched end to a swallow's tail.

Fibrous gypsum forming veinlets with the fibers running crosswise to the vein has a smooth sheen suggestive of a rich fabric, and hence is called "satin spar." All three varieties, alabaster, selenite and satin spar, may and do occur together in some deposits, and merely reflect different degrees of crystallinity and different mechanisms of formation.

In the semi-arid region of south central New Mexico near Alamagordo is the White Sands National Monument, preserving one of nature's wonders: 176,000 acres of dazzling white sand dunes. Many places in the world can boast of larger dunes but it is the mineral that forms the white sands of the Alamagordo desert that makes them unique. It is gypsum—not quartz. From the mountainous area surrounding the Monument in the rainy season, surface waters have dissolved calcium sulfate and carried it downslope beneath the valley fill. During the dry season the water is drawn to the surface by capillary attraction, where it evaporates and the contained calcium sulfate crystallizes as gypsum. The irregular crust formed is broken by wind action into sand grain sizes and whipped into rolling dunes.

Although anhydrite is formed under similar conditions and occurs in deposits as extensive as gypsum, its uses are minor compared with those of its hydrous relative. In fact, anhydrite interferes with gypsum mining if the two minerals are intimately interlaminated. The major use of anhydrite is as a soil conditioner and as a retarder of the setting action of portland cement. However, in West Germany and England it has an interesting and different use. The anhydrite is heated to a high temperature, driving off sulfur dioxide gas, which is then used in the manufacture of sulfuric acid.

Sylvite: Source of Potash

The mineral sylvite, potassium chloride, is a close relative to the more abundant halite, sodium chloride. The two minerals have much in common. They both crystallize in colorless to white cubic crystals, possess a cubic cleavage and have a salty taste. They are both present in sea water although sylvite is much less abundant and does not crystallize until almost all of the water is evaporated. In addition man is dependent for his well-being if not his survival on both of these salts. Halite is the salt for his food and sylvite in large measure makes the food possible. With the ever increasing world population, more food must be grown per acre. To do this on a large scale chemical fertilizers must be used. Potassium compounds are major constituents in such fertilizers and sylvite is the source of the potassium. Approximately ten million tons of sylvite are produced annually, ninety-five per cent of which is consumed by agriculture.

The first identification of potassium in a mineral was in 1797 but compounds of the element had long been known. They were thought to be of organic origin for potash, potassium carbonate, was obtained by leaching wood ashes and evaporating the solution to dryness. Since the process was carried out in iron pots, the name of the element came from "pot" and "ashes." Such was the source of nearly all potassium compounds until 1839 when large salt deposits containing sylvite were discovered at Stassfurt, Germany. Although in the following year the German chemist, Justus von Liebig, pointed out that potassium was one of the elements essential to plant life, major production of potassium did not begin until 1861. From that time on world production of potash has steadily increased to keep pace with the agricultural demands.

Although the use of potash was steadily increasing, the great deposits at Stassfurt, Germany, virtually supplied the world until World War I. At that time the United States was cut off from its supply and an intensive search was undertaken to locate domestic sources. Thus it was that in 1916 a process was devised to extract sylvite from the brines of Searles Lake, California, and production from there continues to the present. Following the war, imports again supplied the major demand and it was not until 1937 that the United States became self-sufficient. This was brought about by the discovery near Carlsbad, New Mexico, at a depth of about one thousand feet, of bedded deposits of sylvite-halite mixture called sylvinite. Since then because of increased production in New Mexico, and new deposits in Utah, the United States has become an exporter of potash.

The world reserves of potassium salts are large. In 1943 sylvite was discovered in extensive deposits in Saskatchewan, Canada, which have been shown to cover a vast area with a total reserve of twenty billion tons. Similar reserves are estimated for the U.S.S.R. Other smaller but major producers are France, Israel and Jordan, Spain and the United Kingdom.

9 Minerals of Land-Locked Lakes

The rocks of the continents and the minerals of which they are made have been subjected since the beginning of the world to weathering and erosion. It is this slow but constant action that has sculptured the surface of the earth into the landscapes with which we are familiar. The minerals, once high on the mountains, are broken grain by grain from the mother rock and gradually work their way down the slopes. Eventually rivers carry them to the sea and their accumulation may build up a thick deposit. Some minerals, of which quartz is the best example, can survive this severe treatment with no change except a diminution of grain size. Most other minerals, either before they are liberated from the host rock or during their long transportation to the sea, are attacked chemically by the waters or gases which surround them, and give rise to new minerals. This process also yields soluble compounds which are carried away in solution.

Since the beginning of geologic time, streams have been discharging their waters containing dissolved salts into the ocean, thus making it continually more salty. A great variety of minerals result when, through various processes, small amounts of these salts separate as solids and settle to the sea floor.

Although most streams eventually find their way to the sea, there are some that discharge their waters into enclosed basins that have no outlet. The Jordan River, flowing south through Palestine, empties into the Dead Sea 1292 feet below sea level, while the Volga River drains into the Caspian Sea ninety-two feet below sea level. Although great volumes of water are continually pouring into these seas, a like amount is evaporated from their surfaces so that their levels remain nearly constant; since water that is evaporated is pure, the salt content of the seas is continuously increasing. In the Caspian Sea the salt content averages nearly thirteen per cent, in the Dead Sea it is about twenty-five per cent, whereas the ocean contains merely 3.3 per cent.

There are many other places in the world where streams have been discharging their waters and dissolved salts into enclosed basins for long periods of time. In the Great Salt Lake in Utah, for example, the percentage of dissolved material is so high that when the temperature falls below 20° F (7° C), salt begins to precipitate. In other areas, where evaporation exceeds the normal inflow of water, a "lake" may be dry except during the rainy season. Such "lakes" in western United States are called playas and are characterized by broad flat surfaces which may be coated with a white crystalline crust. These crusts are composed of a variety of minerals, mostly impure, of which sodium chloride is the principal constituent. All of these salts are at least partly redissolved during the next rainy season. Because of the great variation in the chemistry of different rock types, the minerals crystallizing on evaporation of the water in one region may be considerably different from those in another.

In streams draining areas of recent volcanic activity, there may be, in addition to the soluble products of rock weathering, dissolved material in high concentrations contributed by solfataras and hot springs. The waters from these springs usually contain elements not otherwise present in significant amounts. This is particularly true of the element boron, which is associated with much volcanism, and in places is found emanating from hot springs derived from underlying hot molten granite magmas crystallizing near the earth's surface. The average content of boron in the rocks of the earth's crust is about three parts per million; in the ocean it is somewhat greater, about 4.6 parts per million. Although these are the great reservoirs of boron, the percentages of this element they contain are so minute that they are not a practical source. The lakes of enclosed basins alone have a sufficiently high concentration of boron to yield the suite of borate minerals that are the source of the boron of commerce.

The Borate Minerals

There are about sixty known minerals that contain boron but only a relatively small number are formed through the agency of brines of saline lakes. These are the borates and of these only four, borax, kernite, colemanite, and ulexite, are abundant. Through the years each of these minerals has taken its place as a source of the commercial product borax. The chemical compositions of these borates and some of the less important boron minerals are given below.

Important Boron Minerals

Mineral	Chemical Composition	Mineral	Chemical Composition
Sassolite	H_3BO_3	Priceite	$Ca_4B_{10}O_{19} \cdot 7H_2O$
Kernite	$Na_2B_4O_7 \cdot 4H_2O$	Colemanite	$Ca_2B_6O_{11} \cdot 5H_2O$
Tincalconite	$Na_2B_4O_7 \cdot 5H_2O$	Inyoite	$Ca_2B_6O_{11} \cdot 13H_2O$
Borax	$Na_2B_4O_7 \cdot 10H_2O$	Hydroboracite	$CaMgB_6O_{11} \cdot 6H_2O$
Ulexite	$NaCaB_5O_9 \cdot 8H_2O$	Boracite	$Mg_3B_7O_{13}Cl$

Borax as it occurs in nature is usually impure and because of inclusions of mud has a gray color; when pure it forms in colorless, transparent crystals.

However, once borax is removed from the brine from which it crystallized, it slowly loses part of its water of crystallization and becomes chalky white on the surface and eventually is entirely converted to tincalconite. Thus most of the specimens labeled borax in museums are in reality tincalconite. Also on the dry crusts of desert lakes one usually sees tincalconite rather than borax. Borax is soluble in water and thus to purify the natural product it is necessary only to dissolve it, remove the mud impurities, and recrystallize it.

Kernite, as the formulas show, is similar chemically to borax but with less water. Its outstanding physical property is its easy cleavage in two directions which permits it to break into long splintery fibers. Kernite is less soluble in cold water than is borax but is readily soluble in hot water. Purification of kernite can be accomplished in the same manner as borax, that is, by being dissolved in water and allowed to crystallize. The resulting product, however, is not kernite but borax. This makes kernite unique among industrial minerals since the amount of the finished product exceeds that of the raw material.

Colemanite is frequently found in brilliant multifaced crystals which show one extremely perfect cleavage. When pure it is colorless to white but impurities may render it gray or muddy. Crystals are found in most deposits in cavities in the far more abundant massive colemanite layers and aggregates. One of colemanite's most characterisitic properties is its decrepitation when heated, that is, its tendency to fly apart and form a fine powder. Unlike borax or kernite, colemanite is only slightly soluble in water.

Ulexite is rarely seen in distinct crystals. Its usual occurrence is in loosely packed white fibers aggregated into nodular or rounded masses, that are appropriately known under the name of "cotton balls." In a few rather rare occurrences, ulexite is found in veins, as is asbestos, with closely packed parallel fibers. If such an aggregate is cut and polished at right angles to the length of the fibers, it shows a most interesting phenomenon: light is transmitted in nearly undiminished intensity parallel to the fibers. Thus if such a specimen is placed on a printed page, the printing appears faithfully reproduced on the upper surface. As a result, this variety of ulexite has been called "television spar."

Boron's Many Uses

The list of uses of boron and its compounds is extremely long and becomes longer each year. A small amount of borax is used in fertilizers for soils deficient in boron. It also enters into insecticides and weed-killers. In gasoline it improves the performance of high-compression motors. A great potential use is in high-energy fuels for the propulsion of missiles and aircraft. Boron is used for steel hardening and as a neutron absorber in nuclear reactors.

Boron carbide has a hardness not only far greater than steel but also superior to any natural abrasive with the exception of diamond. It thus finds many uses in drilling bits and other tools. Shortly after the synthesis of diamond under high temperature and high pressure, a form of boron nitride was synthesized in the same manner with crystal structure and hardness comparable to that of diamond.

Some older and more familiar everyday products depend on the mild antiseptic property of borax. These include soaps, detergents and medicine, and pharmaceutical products such as cosmetics, tooth paste, disinfectants and lotions.

One of the principal uses of borax today probably goes back before the beginning of the Christian era. This depends on the property of molten borax to dissolve many metal oxides that otherwise are insoluble. It thus is used as a flux in the soldering, welding and brazing of metals. This was its major use at the beginning of what can be considered the modern borax industry.

The Story of Borax

The first authentic accounts of borax in Europe date from the thirteenth century, when it was introduced from Tibet. Marco Polo visited this region at this time and although he never mentions such a substance, it may have been Marco himself who introduced it to Europe. Little is known of the original Tibetan occurrence of borax but from the fragmentary accounts it appears to have come from enclosed basins in an arid mountain region. There it was found as crusts on dry lakes and in the mud at the bottom as impure crystals called "tincal," a name still in use in some parts of the world today. For six hundred years Tibet was the unchallenged supplier of the world's borax, but its recovery there was primitive and its transportation difficult. One account states that it was carried from the high mountains to the lowland by sheep when they were driven down into the valleys with the coming of cold weather. Whether or not this is true, it required a long overland transport by pack animal to deliver the borax to European markets. Consequently, the price remained high, and enterprising merchants looked for a closer and cheaper source.

It was not until 1827, in Tuscany, in west central Italy, that a new source for boron was discovered. The boron there was in the form of boric acid, the mineral sassolite, obtained from hot aqueous solutions. Natural steam vents had long been known in Tuscany but were avoided as a manifestation of satanic power. In 1777 a chemist, Hubert Hoffer, separated boric acid from the waters resulting from condensed steam. Then in 1827 a French exile, Francesco Larderel, used the heat of the superheated steam to evaporate the water, thus concentrating the boric acid. This launched a chemical industry which today produces not only boric acid but a variety of carbonate and ammonium compounds.

The steam and contained chemicals are considered to be emanations from a large mass of magma crystallizing at a relatively shallow depth below the surface. Had the waters that condensed from the escaping steam accumulated through the centuries in enclosed basins in arid regions, borax and other associated minerals would have been formed. At present the steam issuing from the ground in Tuscany is better known for its use in power generation than for the chemicals recovered from it: more electric power is generated by natural steam in Tuscany than in any other place.

Borax, the normal commercial product, was easily made from boric acid, the long transportation problem was eliminated, and Tuscany soon replaced Tibet as the world's leading producer of boron. But this pre-eminence of Tuscany was later overshadowed by borax from California and Nevada.

The discovery in the western hemisphere was made in 1856 by Dr. John A. Veatch of San Francisco while searching for medicinal spring waters in

(Above) The Ryan Borax Works on the east wall of Death Valley. It was the discovery of colemanite here in 1882 and its mining as a source of boron that ended borate mining on the floor of Death Valley. (Cornelius S. Hurlbut, Jr.)

(Below) A 20-mule team in Death Valley, exact replica of the wagons that carried twenty million pounds of borax out of Death Valley in the 1880's. From the lead mules to the end of the seven-ton water wagon is 150 feet; the total rig weighs forty-seven tons. (Monsanto Magazine)

California. At Tuscan Springs (later named Lick Springs) Dr. Veatch noticed borax crystals in artificially concentrated water. Only a small amount of borax was produced from Tuscan Springs but its presence spurred Dr. Veatch to search further, not for medicinal waters, but for borax itself. Eight months later he visited "Alkali Lake" in the vicinity of Clear Lake as described in a letter written to the Borax Company of California on June 28, 1857:

> "...clambering to the narrow edge of an almost precipitous mountain ridge, we looked down the opposite slope equally steep, on a small muddy lake, that sent up, even to our elevated position, no pleasant perfumes. Thus, on one of the hottest days September ever produced, without a breath of air to dilute the exquisite scent exhaled from two hundred acres of fragrant mud, of an untold depth, I slid down the mountain side into 'Alkali Lake,' waded knee deep into its soapy margin, and filled a bottle with the most diabolical watery compound this side of the Dead Sea. Gathering a few specimens of the matter encrusting the shore, I hastened to escape from a spot very far from being attractive at the time, but which I have since learned to have no prejudice against."

Although the lake (later called Borax Lake) in winter covers about two hundred acres to a depth of three feet, in the dry summer months it extends over only fifty to sixty acres to a depth of but a few inches. During subsequent visits Dr. Veatch probed the mud of the lake bottom to discover a "soapy matter" four feet deep filled with small borax crystals. Beneath this layer a stiff blue clay contained a layer of eighteen inches of larger borax crystals of similar form. A coffer dam three feet square sunk at a point considered to be of medium richness yielded 163 pounds of crystals; a dam in an area presumed to be the poorest yielded 101 pounds. The average of these figures gave a potential production of 127,776,000 pounds for the whole lake. A selling price of fifty cents a pound and a production cost of three cents, yielded a calculated profit of $60,054,720. Production was not begun until 1864, but the output was so great that borax imports into the United States virtually ceased. Most of the natural borax crystals one sees today in museums throughout the world were long ago extracted from the mud of Borax Lake. But after the lake had yielded four years of prosperity for the operators, an artesian well, drilled for experimental purposes, could not be controlled, and the increased volume of water dissolved the crystals and made future operations unprofitable.

About the time Dr. Veatch was wading into Borax Lake, another stagnant pool, Hanchinhama, about four miles to the south, was discovered to contain boron in solution. No crystals were present in the bottom mud but the percentage of boron in the brine was high. On cessation of work at Borax Lake, attention turned to Hanchinhama and for the next six years borax was recovered there by the evaporation of the lake water. Although boron was far from exhausted from the brines at Hanchinhama, exploitation ceased in 1872 with the discovery of another borate mineral, ulexite.

The eastern slope of the Sierra Nevada Mountains marks the beginning of the Basin and Range province that extends eastward through Southern California and Nevada. Here block-faulting has produced numerous ridges with

Ulexite (top) in the unusual form of tightly packed parallel fibers along with chatoyant stones cut from similar material. The great color range of apatite is illustrated by the crystal and cut stones.
(Studio Hartmann)

137

intervening basins. The ridges, some of imposing height, are drained by streams that flow into the enclosed valleys. Because of the arid climate of the region, many of the valley bottoms are occupied by saline lakes. It was from the white crystalline surface of such a "lake" that in 1869 a teamster picked up a ulexite "cotton ball." The specimen eventually reached San Francisco and was identified. When it became known that ulexite as a source of borax could be gathered from the surface of playas, many people began a search for the mineral, and several rich deposits were located in Nevada.

Of the many enclosed depressions east of the Sierra Nevada Mountains where borate minerals have been found, the most celebrated is Death Valley. The discovery of borax there was made by a prospector in 1873, but the white crystalline material he scraped from a lake crust in the bottom of Death Valley attracted little attention. Several years later borax was rediscovered by Aaron Winters who lived with his wife Rosie on the California-Nevada line. He knew of a white crust covering the floor of Death Valley north of Furnace Creek but until a prospector from Nevada told him the story of borax he had considered it worthless. He not only learned how borax occurred but how to test for it: a little sulfuric acid is poured on the powdered mineral, and then a small amount of alcohol is added and ignited. A vivid green flame indicates the presence of boron. Historians have reported the tense moment as Aaron prepared to make the test and his exclamation when he applied the match: "She burns green, Rosie we're rich!"

The Coleman Company began production in 1882 at the Harmony Borax Works, the site of Aaron's discovery, and shortly thereafter at the Amargosa Borax Works to the south of Death Valley. Although also in the desert, this new deposit could be worked when the summer temperatures (130° to 135° F) at Harmony temporarily halted operations.

In the whole story of borax the episode of the "twenty-mule team" is probably the best-known of all. It goes back to a mule skinner, Ed Stiles, the first to handle twenty mules with a 120-foot "jerk line"; he became famous for hauling supplies and water 135 miles from Dagget, California, to the Amargosa Borax Works, and returning with wagons loaded with borax. From the Harmony Borax Works the route was longer and even more difficult. These arduous hauls, which began in 1882, ended abruptly in 1886, not because the "cotton balls" were exhausted, but because a new mineral had been discovered in the mountains.

The Boron Industry

From the time boron minerals were discovered in the United States they were either crystals within the bottom muds of saline lakes or efflorescent products on the lake surfaces. Then a shining white mineral found on the walls of Death Valley south of Furnace Creek wash gave the test for boron. This mineral proved to be unknown to science. The Harmony Borax Works bought the deposit for the Coleman Company and named the new mineral colemanite. Shortly thereafter, Coleman was forced into bankruptcy, and most of his borax properties were bought by F. M. Smith, already known as "Borax" Smith. With the aid of British capital, Smith then organized the Pacific Coast Borax Company which has since dominated the industry in the United States.

Borate minerals: (above right) a splintery cleavage fragment of kernite from Boron, California; (above left) crystals of borax from Salta, Argentina, partially altered to chalky white tincalconite. The individual, hairlike fibers of ulexite visible in the specimen (below right), have been compacted into a "cotton ball" in the specimen (below left). The center specimen is a group of inyoite crystals from Boron, California. (Emil Javorsky)

Several deposits of colemanite were soon located in the mountains, the most significant being at Calico, California, only eight miles from rail transportation. Its immediate exploitation brought a sudden end to production of ulexite "cotton balls" and the long hauls by twenty-mule teams. Colemanite was to remain the chief source of refined borax for forty years.

In 1913, Dr. John Suckow, drilling for water on a homestead claim in the Mojave Desert, struck a layer of colemanite at a depth of 370 feet. Because of the depth, no effort was made to exploit it. However, drilling was later undertaken and in 1925 a sodium borate mineral was brought up from a depth of 380 feet. Not only was a new borate deposit discovered but a new mineral as well—kernite, named after Kern County where it was found. Further exploration showed that the kernite associated with borax lay in a layer one hundred feet thick. All indications pointed to an enormous deposit and the United States Borax Company immediately began underground mining and eventually built a new town, Boron. Thus a great industry sprang up in the middle of the Mojave Desert and signaled the end of colemanite as the source of refined borax.

The Kern County deposit proved to be an oval area about four miles long by one mile wide, the greatest deposit of borate minerals ever found. The ore beds occur at depths of 130 to one thousand feet beneath the surface. Ulexite and colemanite underlie the entire area; the sodium borates, kernite and borax, are found concentrated in three areas and at each of these a mine was located. Underground mining is expensive and may be wasteful, for pillars of ore must be left to support the overlying rock. As a result, after nearly thirty years of working underground, the management decided to go to open pit mining. It was necessary to remove nine million tons of waste material before any borate could be recovered; production began from the open pit in April, 1957. The kernite-borax era is probably the last stage in the borax industry for there is an estimated reserve of 100,000,000 tons of high-grade ore in the deposit at Boron, California.

A visitor to Boron today is struck by the changes in the borax industry since the days when Chinese coolies picked up ulexite "cotton balls" from desert lakes. Now he sees at the bottom of a great open pit men operating shovels from air-conditioned cabs, scooping up borates to be transported to the mill on a conveyor belt.

In tracing the history of the boron industry and the minerals that have been the source of refined borax, one should not omit Searles Lake, California. This lake occupies an enclosed basin into which waters have been draining and evaporating for thousands of years. It differs from other saline lakes in California and Nevada in its great size, nearly forty square miles, and in the large volume of brine that lies below the white crust. Since 1920 the brines of Searles Lake have been used as a source of refined borax. They contain about thirty-five per cent solids of which sodium borate makes up only a small portion. Although the amount of boron in the waters is not great, its extraction from over three million gallons of brine a day makes Searles Lake a major world producer of borax, second only to Boron. These two sources supply ninety per cent of the world's boron.

Borax is but one of the many salts that nature has concentrated in the brines of Searles Lake. Although halite, common salt, is the most abundant, other important chemical products are also recovered. They include: sylvite, potassium chloride, a source of potassium for chemical fertilizers; mirabilite and glauberite, sodium sulfates used in paper processing and in manufacturing household detergents; and trona. Trona is the mineral name of a sodium carbonate abundantly present, and gives its name to the town of Trona on the shore of Searles Lake where the brines are processed. Called soda ash in industry, it finds its largest use in glassmaking; about one ton is used in making five tons of glass.

Wherever surface waters collect in enclosed basins in arid regions, chlorides, sulfates, carbonates and borates can be expected in the resulting brines and efflorescent products. Localities too numerous to mention are found on every continent and their exploitation is a major source of raw materials for the chemical industry.

10 Metals as Minerals

Most minerals are chemical compounds made up of two or more elements, and if one of the constituents is a useful metal, a metallurgical process must be used in extracting it. However, there are twenty elements that occur in nature uncombined with other elements. Of these *native elements*, thirteen are metals. Most of them are extremely rare; only gold, silver, copper and platinum can be considered at all common on earth. Two others, iron and nickel, although rare in the crust of the earth, are common in meteorites, and may be the principal constituents of the earth's core.

Although each of the four most common native metals has properties which distinguish it, all four have a common basic crystal structure. That is, the spherical-like atoms of which they are composed are packed together in similar fashion. To visualize this structure, imagine ball bearings poured into a glass jar and agitated until they settle in place. Thus packed together with minimum space between them, the array is called *closest packing;* portions of the jar will contain a reasonably good model of the structure of these metals. We say "portions of the jar" because, when chance governs, there are two equally possible ways in which this closest packing can be achieved. One way gives rise to crystals with hexagonal symmetry, the other to crystals with cubic symmetry. The latter represents the structure of gold, silver, copper and platinum.

In these metals the electrical forces that bind the atoms together are weak and nondirectional, so that the resulting crystals have a low hardness and show no cleavage. The atoms are, furthermore, so bound that their electrons are free to wander through the crystal structure without allegiance to any particular nucleus; the nuclei may be regarded as fixed and surrounded by a sort of "gas" or fluid of free electrons. As a consequence, all of these metals are excellent conductors of electricity and heat; in fact, silver has the highest conductivity of any material, with copper and gold not far behind.

Although the identity of structure and atomic binding give rise to similar forms of behavior, the native metals differ among themselves in ways dependent on the properties of the individual atoms. The most obvious and easily noted difference is color, which for some elements is so characteristic as to be regarded as standard, as, for example, gold-yellow, silver-white, copper-red. Platinum, however, is similar to silver in color; in fact, it received its name from *plata*, the Spanish word for silver. Gold, silver, and copper have such a strong family resemblance that they are placed together in the *Gold Group* of metals. Platinum on the other hand is unlike them in so many respects that it is placed in another group named the *Platinum Group*, containing six metals of which platinum alone is an important native element. The similarities and differences among the four most common native metals are as follows:

Properties of the Native Metals

Element	Chemical Symbol	Specific Gravity	Hardness	Melting point °C	Boiling point °C
Gold	Au	19.3	2½–3	1063	2600
Silver	Ag	10.5	2½–3	961	1950
Copper	Cu	8.9	2½–3	1083	2336
Platinum	Pt	21.5	4–4½	1774	4300

In well-crystallized specimens one sees another similarity between gold, silver, and copper. Isolated single crystals of any of them are a great rarity but they all occur in delicately reticulated and arborescent growths. Moreover, some crystals of each mineral are found as slender wires elongated in a definite crystallographic direction; whereas others are thin plates flattened parallel to a single crystal surface. Platinum on the other hand crystallizes in no such pleasing and attractive shapes and when the rare crystal is found, it usually is a misshaped cube.

Man's First Metal: Gold

Early in the morning of January 24, 1848, James Marshall walked to the tailrace of the dam he was constructing for a sawmill owned by John "Captain" Sutter near Sacramento, California. Seeing a golden glint in the shallow water, Marshall reached down and picked up a piece of metal. Pounding it on a flat stone, he discovered it was easily malleable and flattened out into a little plate. Next morning, more metal had accumulated in the tailrace, and a frenzy of testing ensued. The suspect metal was bitten, pounded on an anvil, boiled in vinegar, and weighed against silver on a scale. When balanced against an equal weight of silver in air, the metal outweighed the silver under water: it was gold! There was little excitement at first, but when in May 1848 Samuel Brannan paraded through the streets of San Francisco with a bottle of gold dust, shouting "Gold! Gold! Gold from the American River!", thousands took up the cry. It is estimated that ten million dollars in gold was taken from California's streams in the first year following discovery.

The miners found their gold at shallow depths in the river beds, sometimes as "pay streaks" in the gravels, more often in "riffles" or transverse cracks in the bedrock over which the stream flowed. This was placer gold washed from lode deposits high in the mountains and concentrated within easy reach of prospectors. Gold is chemically inert, malleable, and plastic, and extremely heavy. Hence when nature ground to powder the brittle minerals in which it was embedded, the gold was merely pounded into flattened nuggets and flakes. By virtue of their high density, they worked their way through the sands and gravels to the stream bottom to lodge behind obstructions and in cracks in the bedrock. From these cracks the early Forty-niners dug out nuggets of gold, often with their pocketknives. The first mining equipment was a pan into which the miner scooped sand and water, and by a swirling motion washed the lighter minerals over the edge, leaving only a small quantity of heavier minerals. Among them might be flakes of gold, "colors," or an occasional larger nugget. When the richest placers were exhausted, the miners took to the harder work of washing the gravels in sluice boxes, which caught the gold behind artificial riffles as the running water swept the lighter minerals away. Diverting the stream by damming and digging down through several feet of gravel, they worked on the exposed bedrock with slender "sniper's picks" and long, thin spoons. As the richer streaks were exhausted, the sluices were improved by putting mercury (quicksilver) in the crevices behind the riffles, to catch by amalgamation the very finest gold particles that might otherwise be swept over the riffle and lost. The gold was then recovered by distilling off the mercury, which was caught in a retort and returned almost undiminished to the riffles. With the exception of amalgamating, there was nothing the Forty-niners did to win gold that an Anatolian or Achaean of the third millenium B.C. could not have done.

Glistening white piles of crushed quartz from which gold has been extracted stretch for scores of miles east and west of Johannesburg, South Africa, since 1886 the world's chief supplier of gold. (Cornelius S. Hurlbut, Jr.)

Gold is sought as eagerly today as in A.D. 1848 or 1848 B.C. and the demand still exceeds the supply. In some countries, such as the United States and Great Britain, the possession of gold is a government monopoly and its price is fixed by law. In many other countries, however, gold is sold as a commodity on a free market with a fluctuating price subject to supply and demand, and it is significant that in recent times the free market has always stood high above the legally fixed price. Part of the value of gold lies today in its acceptance as a monetary standard, but men of all ages have sought, died for, and even worshipped gold. Its unique, inimitable and pleasing color was always a part of its appeal. Sun-worshippers saw in it the congealed quintessence of their deity—sunlight made tangible. Alchemists argued that all other metals were corrupt since they could be altered chemically, and only gold incorruptible and divine; hence, if a metal could be made to die and then be reborn cleansed of its taint, gold must be the product. To test the weight of a large gold nugget in the hand, and feel the smooth, cool mass, is a most satisfying experience. Further, gold is the ideal metal for craftsmen. It casts well, may be hammered into the thinnest sheets of any metal (as thin as $1/_{100,000}$ of an inch), may be drawn readily into wire, carved, buffed to a glowing polish, heated repeatedly without discoloring, and joined to itself and other metals by hard soldering without the use of fluxes.

The color of natural gold varies from specimen to specimen; some are a deep rich yellow; others are much paler. The color difference depends on the amount of silver in the metal. Silver also lowers the specific gravity in simple proportion to the amount present. A pale-yellow specimen of native gold from the Don Basin in Russia has a specific gravity of 16.9, and analysis reveals the composition to be 85.21 per cent gold, 14.71 per cent silver. To a much lesser extent, copper is present in native gold and also diminishes its specific gravity; but imparts to it a ruddy hue.

The reason for the frequent presence of appreciable silver in gold is that the atomic structure of silver and gold is the same and the size of the atoms nearly identical. Thus, during the process of crystallization, some silver atoms can take the place of gold atoms without disrupting the structure. In fact, silver and gold atoms can substitute for each other in any proportion, so that it is possible to find natural alloys of silver and gold in which the amounts of the two metals are approximately equal. The reason for the small amount of copper generally found in natural gold is the smaller size of the copper atom, which makes it a poor "fit" for gold, although the crystal structure of the two metals is the same.

Gold in Prehistory

The yellow color of gold is not only pleasing but tends to catch the eye, making it possible to detect small grains in an aggregate of other minerals. The tiniest flakes, of which thousands are required to make an ounce, are easily visible to the prospector. For this reason one can understand why gold was found at such an early date by man. Even in Neolithic graves, flakes and nuggets of gold have been found, suggesting that it was valued as an ornament or amulet. Exquisitely worked golden ornaments occur in the so-called Royal Graves of Ur, in approximately 2700–2600 B.C. As early as to the fourth millenium B.C.

golden objects appeared in the culture of the Nile Valley but showed little artistry or mastery of metal skills. The first record of gold mining in Egypt dates from the Sixth Dynasty, about 2200 B.C., and the art of fashioning golden objects had reached a high degree of perfection by 1350 B.C. In that year the boyking Tutankhamen was buried in an inner coffin of solid gold weighing 242 pounds! This massive piece of metal was cast and then shaped by tooling into a finely detailed portrait of the king. Buried with one of Egypt's least important Pharaohs, this coffin indicates a mastery of metal working unexcelled in any age, and a prodigality with gold almost unmatched in history.

Gold has always been considered one of the spoils of war, and much of the precious metal in existence during early time passed from one conqueror to another. The gold loot in Alexander the Great's conquest of Egypt alone was reported worth over $100,000,000. Although the Romans mined gold in various places in Europe, much of their early store was plunder that had already passed from Egyptians to Greeks.

Much later, in the sixteenth century, the Spaniards found gold in the New World first as accumulated treasure and later in the ground. From the Aztecs in Mexico, Cortez in 1519–1521 plundered great stores of gold, and in 1524 to 1533 Pizarro wrested from the Incas a nearly equal amount. As a direct result of the Spanish conquest of America, the gold in Europe increased during the sixteenth century from 160 tons to over six hundred tons.

The Spaniards, in their lust for gold from America, shipped to Europe several boatloads of the iron sulfide, pyrite, a brassy, metallic mineral. They thus learned the truth of "all that glitters is not gold." Pyrite, or "fool's gold," is but one of the yellow minerals that through the centuries has been mistaken for gold. Others are chalcopyrite, a soft yellow copper mineral, and vermiculite, a glittering yellow mica. The identification of gold can easily be checked. The gold of jewelry is alloyed with other metals to make it hard and durable; by comparison native gold is soft and can easily be scratched. Thus to test for gold one can try scratching the mineral with a knife; if it leaves a smooth shining furrow, it is gold. Pyrite can not be so scratched; chalcopyrite can be scratched but leaves a greenish black powder, and vermiculite splits apart into thin flakes.

Since gold is malleable, another test is to place the mineral on an anvil and strike it with a hammer. If it is gold, it will flatten out, whereas any other yellow mineral will shatter.

Prospectors learned long ago that "gold is where you find it." It is found on every continent and is mined in high mountains and in deserts, beneath the permanently frozen ground of the arctic and beneath the deeply weathered soil of the tropics. However, there are some places that are more likely to reward the prospector than others, but these are determined by geological environment rather than by geographical location. Gold is associated with rocks of all ages and types, but it is more likely to be found in deposits related to granitic rather than to gabbroic rocks. Although there are many types of gold deposits, the most universal is gold-quartz veins found in the neighborhood of granitic rocks and probably genetically related to granite.

In such quartz veins gold is usually accompanied by the far more abundant pyrite. Under the conditions that exist at the surface of the earth, pyrite is very unstable and breaks down, yielding limonite, a yellow-brown iron oxide. Although the presence of limonite does not guarantee the presence of gold,

(Above) A large nugget of gold rounded and polished by the sands of a stream.

(Below left) Flattened octahedral crystals of gold join in a dendritic pattern to form a golden fan. (Both by Benjamin M. Shaub)

(Below right) Wires and plates of gold in a matrix of quartz. (Tozier Collection, Harvard University)

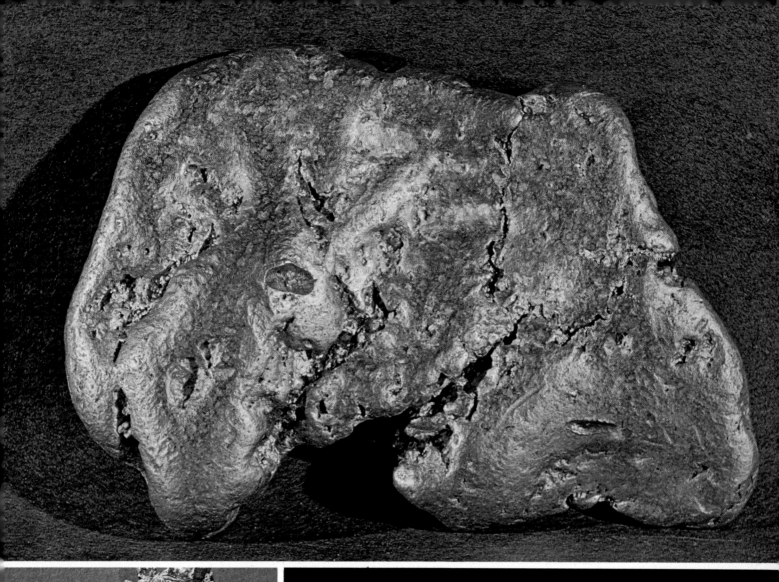
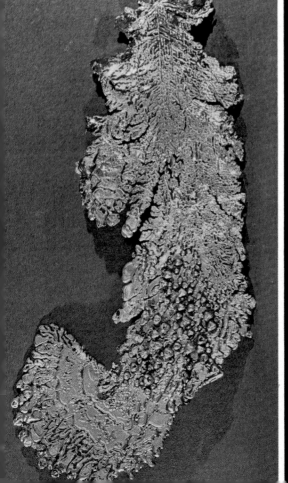
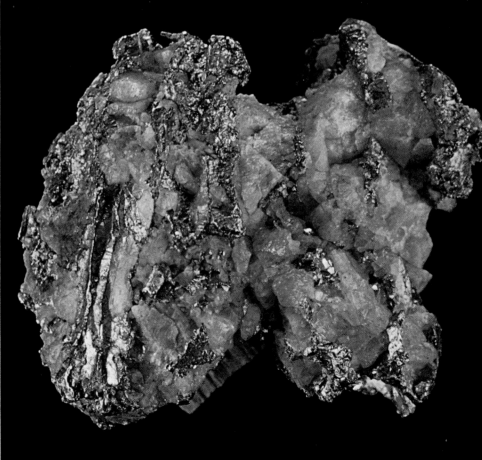

the iron-stained outcrops called *gossans* are considered a guide to ore.

Gold is a "noble" metal, that is, it does not react in the laboratory with any of the ordinary acids or alkalies or in nature with the relatively dilute chemical reagents present in the soil and rocks. While the rocks enclosing the quartz veins are slowly altered by chemical weathering processes and the veins themselves are fragmented by mechanical weathering, the gold remains unchanged. As rock minerals are gradually worn away, the gold particles may remain essentially in place, building up in an eluvial deposit a concentration of gold. It is recorded that about 2000 B.C. such a deposit, covering an area of one hundred square miles, was worked in Nubia to a depth of seven feet and produced an estimated one thousand tons of gold. An eluvial deposit in which the gold remains behind differs from a placer deposit in which the liberated gold particles are washed into the neighboring stream.

However, placer and eluvial mining have not been confined to primitive people, for nearly every major gold deposit discovered during historic time was first located as a surface or stream deposit. When gold is mined from a stream, it is the dream of every miner to trace it headward and find its source, the mother lode, as gold-bearing quartz veins in the mountains.

The Great Gold Rushes

It was nuggets of placer gold found at Coloma, California, on the south fork of the American River, that signaled the gold rush of '49, the greatest of all time. During the next thirteen years, placers yielded nearly $650,000,000 in gold. California remained the foremost gold producer in the United States for one hundred years but after the first easy gold was won from the placers, most of the production came from the Mother Lode, the name given to the zone of gold-quartz veins that extend north-south for two hundred miles along the western slope of the Sierra Nevada.

In 1851, only two years after the gold rush to California, gold was discovered in New South Wales, Australia, and shortly thereafter in nearby Victoria. The development of Australian gold mining closely paralleled that in California. A tremendous rush followed the discovery of placer gold. Prospectors who came by the shipload were frequently joined by the ship's crew, who deserted the ship, lured by the hope of quick riches in the goldfields. With the exhaustion of the alluvial gold, the individual digger gave way to the corporation with the necessary capital to mine the gold-quartz veins in the hard rock.

Another parallel between the United States and Australia was the discovery toward the end of the nineteenth century of additional rich deposits in both countries: at Cripple Creek, Colorado, and at Kalgoorlie in Western Australia. These two discoveries were not of native gold but gold tellurides, calaverite and sylvanite. Over $400,000,000 in gold has been taken from each of these districts.

South African Gold

Spectacular as the gold rushes were and as impressive as were their yields, they are completely overshadowed by the production of gold in the Transvaal, Republic of South Africa. Although gold had been reported earlier, it was in

(Above) Native copper from Keweenaw peninsula, Michigan, imparts a reddish hue to a nodule of massive datolite and a crystal of transparent calcite in which it is incorporated.

(Below left) Ruby-red crystals of cuprite and crystals of native copper from Tsumeb, Southwest Africa, have formed by the oxidation of copper sulfide minerals. (Both by Emil Javorsky)

(Below right) Native copper crystals from the Keweenaw peninsula, Michigan, where millions of tons of copper were mined as the pure native metal during the second half of the nineteenth century. (From the Ted Boente collection. Photo by John H. Gerard)

1886 on the Witwatersrand, near the present city of Johannesburg, that it was discovered. Almost immediately it was recognized as the world's major gold deposit.

Unlike the normal occurrence of gold in gold-quartz veins, here it was found in an ancient conglomerate, with the gold concentrated in narrow sheets, called reefs. The South African geologists believe it to be a gigantic placer deposit in which the pebbles of the conglomerate and the particles of gold were contributed from a pre-Cambrian mountain range. The conglomerates of the Witwatersrand series extending east-west for 120 miles and dipping to the south have yielded gold over their entire length. As mining follows the gold-bearing reefs down the dip to ever increasing depths, the problems of mining become greater. The temperature increases about 1° F for every 180 feet of depth, so today at depths well over ten thousand feet, the temperature is 120–125° F in the deep mines. This necessitates elaborate ventilating and cooling systems, the cost of which increases with depth and may set an economic limit to the mining even though gold is still present.

In spite of difficulties, the production from this most remarkable of all gold fields has continued to increase; it produces over forty per cent of the annual gold production of the world. Since its beginning the Rand has produced over eighteen thousand tons of gold valued at about eighteen billion dollars! Nor is the end of South African gold in sight. In the Orange Free State, 150 miles to the south, geophysical prospecting located far below the surface what is believed to be the southern extension of the Witwatersrand conglomerate. Whether or not the conglomerate extends over this entire distance, drilling to a depth of five thousand feet has penetrated a gold-bearing reef. From shafts sunk to that depth are now coming gold ores which may rival if not surpass the riches obtained from their northern counterpart.

Silver—the White Gold

It is surprising to learn that silver, today worth only one-twenty-fifth to one-fifteenth as much as gold, was valued more highly than gold by many ancient civilizations. This was true of the Sumerians, Babylonians and Hittites, who used it as the basis of their monetary systems. In Egypt in early Dynastic times silver, called "white gold," was regarded as scarcer and hence more valuable than gold. Indeed, for a time, even copper was valued more highly than gold in Egypt. However, the inflow of smelted copper and silver into Egypt after about 2500 B.C. reduced the values of these metals relative to gold. The high regard in which silver was held resulted partly from its scarcity as a native metal, and partly because of its religious significance. Silver is as naturally the metal of the moon as gold is of the sun, and the chief deities of Sumer and Babylonia included a Moon-god and goddess, but not a Sun-god. The "silver" of these ancient kingdoms was, of course, not pure silver, but was liberally alloyed with gold and copper indicating its origin as a native metal. Thus, the rein-rings from the so-called Royal Tombs of Ur contained 93.5% silver, 6.1% copper and 0.4% gold. A "silver" object of the Pyramid Age of Egypt contained 90.1% silver, 8.9% gold and 1.0% copper.

Silver is chemically more reactive than gold and forms several natural compounds which are its chief ore minerals. Further, even when native, silver is always deeply tarnished and is not as readily recognizable in the outcrop as is gold. This further accounts for the relative scarcity of silver in ancient times.

In only a few of its widespread occurrences has native silver been an important ore. The occurrences can be divided into those in which native silver is primary and those in which it is secondary, formed by the alteration of earlier primary minerals. From the commercial point of view the first is more important, but the second type is found in more localities.

When minerals containing silver are subjected to weathering at the earth's surface, chemical reactions may take place, producing secondary silver minerals. Chief among these are silver chloride, the mineral cerargyrite or "horn silver," and native silver. In some localities, these two minerals have constituted rich ores in the early stages of mining, but they do not persist in depth, and give way shortly to primary silver minerals, which may be much leaner ores. Thus, the American southwest is studded with "ghost towns" or abandoned mining camps called "Chloride," mute testimony to the short-lived "boom-and-bust" history of mining ventures dependent on secondary silver ores. At most localities where silver has been mined, some native silver has been found as a surficial product, but no great mining enterprise has ever been based on silver ores of this type. This is undoubtedly the type of native silver found and used by ancient peoples.

Native silver as a primary mineral is of greater significance to the mining industry. At Kongsberg, Norway, it occurs in extensive low-temperature veins, associated with zeolites. Here native silver was discovered in 1623 and was mined continuously for over three hundred years. Large masses of pure silver, one weighing over fifteen hundred pounds, and many of the most beautiful specimens of crystallized silver seen in museums have come from Kongsberg.

Long before silver was known in Norway it was being mined in Saxony, in what is now East Germany. About 1163, native silver was discovered near Freiberg in veins associated with arsenides and sulfides of cobalt and nickel and with native bismuth. Similar deposits of native silver were later found both north and south of the Erzgebirge "ore mountains," which separates eastern Germany from Czechoslovakia. Schneeberg and Annaberg in Saxony and, to the south, Joachimsthal in Bohemia became the most famous producers of silver. The discovery of silver at Schneeberg was not made until 1470; but so rich were the veins and so feverish the excitement attending the discovery that within four years there were 176 producing mines. Activity was short-lived for by 1500 most of the silver had been worked out to a depth of one thousand feet.

The minerals associated with silver in the veins of the Erzgebirge formed a characteristic assemblage. They included niccolite, skutterudite, and cobaltite, arsenides of nickel, iron and cobalt; and native bismuth. The medieval Saxon miners endeavored unsuccessfully to smelt these minerals. Niccolite is a copper-red mineral, but since no copper could be extracted from it, it was called *kupfer nickel*, "nickel" referring to the belief that it had been bewitched by those evil underground spirits, the Nixes, just as the Cobolds, who also dwelt underground had rendered the silver-colored skutterudite and cobaltite worthless as silver ores. Thus, the names *nickel* and *cobalt* given to these elements centuries later had their origin in a Saxon superstition.

While the mines at Schneeberg were still active, native silver with the same mineral association was found twenty miles away at Annaberg in 1492. Mining

activity here lasted for about one hundred years during which time 969 mines were operated. It was at this time that Georgius Agricola was the town physician and inspector of mines at Joachimsthal, and it was about the mining practices in the Erzgebirge that his books, landmarks in mineralogy, were written.

In several of these deposits, uranium minerals are also present. Although alteration products of the uranium minerals give the outcrops a green or yellow color, uranium is more abundant in depth as the primary mineral, uraninite or "pitchblende." In general, there is a vertical zoning in these deposits with silver most abundant at the surface, cobalt, nickel and bismuth minerals predominating in the middle workings, and uranium in the lower levels. With the discovery of radium in 1898, the uranium minerals, formerly merely mineralogical curiosities, became a source of that element and brought about a renewed interest in the district, particularly in Joachimsthal, which furnished the material from which the Curies isolated the new element. Joachimsthal now lies behind the iron curtain so that little is known of recent mining activity there; however, it is believed that since the advent of atomic weapons, it has become an important source of uranium.

Minerals are known by the company they keep, and no mineral association is more characteristic than that of the nickel-cobalt arsenides with native silver.

In the spring of 1903, a construction worker on the Tmiskaming and Northern Ontario Railway picked up a copper-red mineral near the present town of

(Above left) Each feather-like protuberance in this crystal of native silver from Kongsberg, Norway, joins the central stem at the same angle, which indicates that they all belong to the same crystal.

(Above right) Native silver tends to crystallize as wires (as in this specimen from Channarcillo, Chile) parallel to a definite crystallographic direction but may later bend and curve. (Both by Wilbert Draisin)

Cobalt, Ontario. Since no one in the construction party was a geologist or mineralogist, the find aroused little interest. Later that year, W. G. Miller, the Provincial Geologist, identified the specimen as niccolite, a valuable ore, but knowing the mineral association in Saxony, suspected that it also might indicate the presence of other minerals including native silver. By the time Miller visited the district in November, four veins rich in native silver had already been located. But it was not until eighteen months later, when shipments of ore made prospectors aware of the riches of Cobalt, that major interest was shown. The rush to stake claims was followed by an unparalleled production of silver. In the ten years following the discovery approximately $100,000,000 of silver had been taken from the veins at Cobalt.

Twenty-seven years later, while prospecting from the air in Canada's Northwest Territories, Gilbert LaBine noticed a bright-colored outcrop near the eastern end of Great Bear Lake. After landing on the lake for closer investigation, he discovered native silver and cobalt minerals on a small island a short distance from shore. Later, on the mainland he found four major, richly mineralized veins containing not only silver but cobalt, nickel and uranium minerals. Although in 1930 Great Bear Lake was in a wilderness nine hundred miles away from road or railroad, the richness of the deposit encouraged immediate exploitation. For most of the year the only transportation was by air, supplies being flown in and ore flown out. At first silver alone was recovered. Later pitchblende (uraninite) was mined, first as a source of radium and then for uranium as a source of atomic energy.

Copper in Antiquity

Although the importance of native metals in the development of human technology can scarcely be questioned, there is little likelihood that the precise details of this important aspect of human prehistory will ever be known. The earliest appearance of worked metal was at Catal Hüyük in Turkey, where little tubes hammered from native copper were found in 1962, decorating the fringes of string skirts. The date of the Catal Hüyük culture, still very uncertain, is probably between 6000 and 5500 B.C.

For an understanding of the role played by native copper in such very early cultures, we can turn to a culture half a world away that has survived nearly to our own time. When, in 1776, an English adventurer, Samuel Hearne, penetrated the wilderness northwest of Hudson's Bay, he found a tribe of Eskimos skilled in making fish hooks, ice chisels, gaff hooks and weapons from lumps of pure copper found in the glacial debris. In 1919 an anthropologist visited these "Copper Eskimos" and learned how they fashioned their tools and weapons. Their smiths hammered out the irregular nuggets on a beach pebble held as an anvil in the hand. After rough shaping in this way, the object was finished by rubbing and grinding on a large stone. Although nearly 150 years had passed since Hearne's visit, this highly conservative culture had not significantly modified their treatment of the copper nuggets.

Neolithic communities, like those of Anatolia in the sixth millenium B.C., had no effective way of mining or cutting large masses of copper, even when such masses were available. Consequently they were restricted, like the Copper Eskimos, to small masses, which they probably worked by a cold-hammering

Layers of black chromite alternating with anorthosite at Dwars River in the eastern Transvaal, South Africa. (H. Kruparz: Geological Survey of South Africa)

technique. This method is not very satisfactory, since the copper becomes hardened and rather brittle after only a moderate amount of hammering. The hardening imparts more durable qualities to the finished tool, but the object being fashioned is likely to crack unless annealed by periodic heating. There is a good probability that the discovery of annealing and hence of hot working, melting, and casting took place before smelting of ores was discovered. The finds of metal hoards in Iran at Sialk, in Iraq at Tell Halaf, and in Egypt, indicate the prevalence during the fourth millenium B.C. of melting and casting of gold, silver and copper.

There is a great gulf between the simple process of rendering liquid a metal that otherwise retains its identifiable characteristics and the magical transmutation of a green, blue or black ore mineral by fire into metal. Thus before man could win metal from those ore minerals there was a long interval, perhaps of thousands of years duration, in which native metals, chiefly copper, furnished man's best tools and most cherished adornments. Not all cultures attained the art of smelting at the same time. If a survey could have been made in 1000 B.C. it would have found that some people like the Hittites had entered the Iron Age; others like the Greeks were in the Bronze Age; whereas many were in a condition of Neolithic barbarism. Indeed, some otherwise highly developed cultures never did attain the arts of metallurgy: for them native metals continued to suffice. The outstanding examples of this arrested development are the cultures of the American Indians.

Indians of the Lake Superior region learned to use and value the copper found in abundance on Isle Royale and on the Keweenaw peninsula at an early date. A typical Indian mine on Isle Royale was a trench some two hundred feet in length and as much as twenty-five feet deep from which chunks of copper were won by primitive methods. Fires were burned against the face of the rock and then cold water cast on the heated rock to crack it, after which the loosened material was dug and battered away with stone mauls. This copper was widely traded and artifacts made from it have been found at such distant points as Oklahoma, Georgia and the Pacific coast. Great Lakes copper is recognizable in analysis by its high purity and silver content, and can be distinguished readily from Mexican or South American copper, which contains arsenic and other impurities. Among these artifacts are copper "breastplates" or ornamental sheets, sometimes elaborately embossed with human or animal designs. Knives, arrowheads, fishhooks, earplugs, and votive or ceremonial objects were made from Great Lakes copper and are found in Indian burials throughout the central and southern United States. Although these objects are beautifully finished and show great skill in cold-working, there is no evidence that North American Indians ever learned to melt and cast metals like the Indians of Peru. It is probable that the Peruvian Incas were in the early stages of a true Bronze Age when their culture was extinguished by Spanish conquest in the sixteenth century.

It is interesting to note that in the recently excavated tombs of the kings and queens of the First Dynasty of Egypt, about 3000 B.C., copper tools were found, but no bronze, and no evidence of extractive metallurgy. The manicure set of Hetep-Heres, a queen of the Sixth Dynasty, was of copper with ivory handles. This adherence to copper, much of it no doubt native copper imported from Asia Minor or Cyprus, was partly the result of backwardness in technology, but perhaps also of traditional Egyptian conservatism. We will probably never learn from the plundered tombs of Egypt's kings the secrets of metallurgy of the humble artisans whose bones lie in unmarked graves in Egypt's sands.

Native Copper

Most of the copper produced today comes from compounds in which it is combined with other elements. Compared with these, native copper is a rarity. However, it is widely distributed in small amounts in the oxidized zones of copper deposits associated with azurite, malachite and cuprite. No large masses of copper have been formed in this way but in all probability it was this type of native copper that was first found and used by early man. Other deposits in which the native copper appears to be primary rather than secondary are associated with intrusive rocks of basaltic composition. These deposits are also small and the copper from them is a mineralogical curiosity rather than an economic source. In only two places in the world has native copper been found abundantly enough to be considered an ore. One is at Coro-Coro, Bolivia, where disseminated native copper is found in a sandstone. The other, and far more important, is on Keweenaw peninsula in Lake Superior. Shortly after founding the city of Quebec in 1608, the French explorer Champlain was presented by the Algonquins with a chunk of solid copper said to have come from the bank of "a great river which flowed into a great lake." During the following two hundred years white settlers in the midwest found Indian artifacts of copper and heard Indian legends of the rich copper country somewhere on the south shore of Lake Superior. It was not until the summer of 1840 that an explorer, Douglass Houghton, with a small party, explored Michigan's Upper Peninsula and returned with proof of an abundance of copper there.

Houghton's report brought a rush of prospectors who knew little of the country and less of mining. Since the geology was unknown, once the miner's shaft struck barren rock he did not know which way to turn to find more ore.

When, after a decade, the geology became known, it was shown that the Keweenaw peninsula for the two hundred miles from northern tip to southern base was composed of a series of very ancient conglomerates and lava flows dipping to the northwest beneath Lake Superior to emerge again on Isle Royale. It was through the porous tops of the lava flows and around the pebbles of the conglomerates that rising hot solutions had precipitated the native copper and a small amount of native silver. The solutions also found their way into cross-cutting fractures, and it was copper deposited there that attracted the early prospectors and determined the location of the first mines. It was later learned that although the conglomerates and lavas carried less copper per ton than the cross-cutting veins, they were far more extensive. Thus, the great mines of the Copper Range followed the exposed edges and down the dip of the lava flows and conglomerates of the Keweenaw Series.

Although most of the copper occurred in irregular masses filling cracks and replacing earlier minerals, it was also found as the cementing material between rounded pebbles of the conglomerate. In some places the pebbles themselves were completely replaced. Although most of the copper masses were small, some were so large that they presented difficulties in mining. Native copper is tough and difficult to break, and being malleable, is nearly impossible to drill with normal mining equipment. The largest mass encountered underground was $44 \times 22 \times 8$ feet. This and other similar large masses had to be cut or sawed into small pieces before they could be removed from the mine.

The mines in the "Copper Country" are now closed but their one hundred years of productive life left their impact on American mining. Keweenaw peninsula saw the first mining boom in the United States and became its first great copper-producing district. "Lake Copper," because of the absence of impurities of arsenic and sulfur, became a standard of purity. Engineering triumphs included sinking the first shaft to a depth of over one mile, and the development of steam hoists and portable compressors for air drills. During the lifetime of the mines, over $325,000,000 was paid in dividends, but their legacy is not measured in dollars alone. It includes the Michigan College of Mines, established in 1882 at Houghton, as well as generations of practical miners. The men trained in both the college and mines have gone out to mining camps around the world, taking with them skills learned on the Copper Range.

The mines also left a legacy to mineralogy. Sixty-seven minerals have been described from them. Although a large number of these are rock-forming minerals and uninteresting to the collector, many a mineral collection is enriched and a museum display enhanced by the presence of the more unusual Michigan minerals. The copper itself in well-crystallized and arborescent forms is of course the best-known and the most important. Highly valued are specimens of native copper known as "half breed" which contain irregular masses or crystals of native silver. Zeolites, apophyllite, prehnite, epidote, and datolite have been found in fine crystals, mostly in cavities in the lava flows. Specimens unique to the Copper Country are clear transparent calcite crystals enclosing delicate crystals of copper; and fine-grained datolite incorporating finely divided native copper and made red by its presence.

The Platinum Group

Although platinum occurs in the native state, there is no evidence that primitive man used it as he used gold, silver and copper. This is probably because it never occurs in such large masses as the three other metals and the small silvery grains in placers do not, like tiny flakes of gold, attract attention. Further, the melting point is so high that it would have been impossible to melt together many small fragments to obtain a mass large enough for working. Platinum is a young metal: the first mention of it occurred in 1741 when it was referred to as a *"platina del Pinto"* coming in grains and nuggets from New Granada, now Colombia. *Platina* is a diminutive of *plata,* Spanish for silver, the name being undoubtedly given to the metal in allusion to its silver color. Rio di Pinto is the name of the stream, a tributary of the San Juan, in whose gravels it was found as a placer mineral.

In nature, platinum is never pure but is alloyed with other metals of the platinum group and usually contains considerable iron and small amounts of copper, gold and nickel. Of the six metals listed in the table below, platinum is by far the most abundant.

Platinum Group Metals

Heavy		*Light*	
Metal	*Specific Gravity*	*Metal*	*Specific Gravity*
Platinum	21.45	Palladium	11.90
Isidium	22.38	Rhodium	12.10
Osmium	22.47	Ruthenium	12.26

Platinum, which today has so many uses in the arts and sciences and is associated with objects of great value, had a lowly beginning. It was known that the specific gravity of gold (19.3) would be lessened by the addition of any other metal and that adulteration could be thus detected. However, if platinum with a specific gravity equal to or greater than that of gold were added to gold, it could not be detected. As a result, it was reported that late in the eighteenth century the King of Spain ordered the platina mines closed to prevent the fraudulent adulteration of gold.

The stream gravels of Colombia remained the only source of platinum from the time of its discovery until 1824. In that year prospectors found native platinum on the eastern slope of the Ural Mountains in Russia while working for placer gold.

Although Colombia continued production, Russia almost immediately became the world's major producer, a position it held for over one hundred years. For a short while Russia, unable to dispose of its newly found metal, used it for coinage. However, as new uses were found for platinum, it gradually increased in value, and by 1835 the bullion value of the metal exceeded the face value of the coins and they immediately went out of circulation.

Attempts to trace the platinum from the stream gravels to its source led the prospectors west into the Ural Mountains, in the neighborhood of Nizhni Tagil. No obvious veins containing platinum could be found but the trails always ended in an area underlain by a dark rock called *dunite,* made up mostly of olivine and frequently altered to serpentine. Chromite is present in irregular masses in the dunite and serpentine; it was in the dunite, frequently associated with the chromite, that platinum was found in place, but in such small amounts that attempts to mine it in place were largely unsuccessful. It remained more profitable to wash it from the stream beds where nature had been concentrating it for tens of thousands of years.

The figures on the world production of platinum are most incomplete and inaccurate. Before the revolution in Russia, production figures were frequently falsified to avoid payment of taxes; since the revolution, no official figures have been available. Nevertheless, it seems likely that in 1935 Canadian production of platinum exceeded the Russian. In Canada, platinum was found not as the native metal but in the mineral sperrylite, platinum arsenide, associated with the copper-nickel ores at Sudbury, Ontario.

The Great South African Platinum Deposits

The most recent major finds of native platinum have been in South Africa: in 1923 in the Waterberg district and a year later in the Bushveld Igneous Complex. In 1929, Percy A. Wagner in the preface to his book *The Platinum Deposits and Mines of South Africa* made the prophetic statement: "The Transvaal, that marvelous storehouse of mineral wealth, has become of recent years, an important producer of the platinum metals. She is destined to become the world's leading producer as her primary deposits of these metals are incomparably the greatest." By 1953, South Africa had become the world's leading producer of platinum.

The Bushveld Igneous Complex is one of the outstanding geological features of the world. It occupies an elliptical area in the Transvaal roughly three hundred miles from east to west and 120 miles from north to south. It is a complex of igneous rocks varying in composition from peridotite to granite. It is a sheetlike body of gigantic size that conforms to the inward-dipping, basin-like form of the underlying strata, the Magaliesberg quartzite. The igneous body itself is layered, with the layers in general paralleling the basal contact. The traditional view is that the bedlike layering of the many different rock types is the result of the settling out of successive crops of crystals from a single parent magma. Although no layer can be traced completely around the body, many with uniform thickness and dip can be followed for scores of miles. It was in such a persistent horizon that Dr. Hans Merensky discovered platinum in 1924.

The platinum-bearing layer now called the *Merensky Reef* is a thin sheet conforming to the pseudo-stratification of the enclosing rocks, and composed of norite, chromite and pyroxenite. The platinum occurs mostly as the native metal but some sperrylite, as well as chalcopyrite, pentlandite and pyrrhotite, is present. It is from this horizon near Rustenburg that 400,000 ounces of platinum are produced each year. From the striking mineralogical similarities among the major producing localities, it may be safely concluded that platinum genetically is associated with dark igneous rocks rich in olivine and pyroxene and with minerals of copper, iron, nickel and chromium.

11 Ore Minerals of the Common Metals

With the advance in metallurgical skills special uses have been found for many metals that were only curiosities a generation ago. Some of these so-called "exotic" metals occur only as "impurities" and are recovered from ores of other metals. Thus gallium, used in dental alloys, is a by-product of aluminum production; and rhenium, used in corrosion-resistant alloys, is found only in molybdenum minerals. Since such elements are rare, their role in the age of metals will always be limited. However, this century has seen not only the development of small uses of rare metals but large uses of two of the commonest metals, aluminum and magnesium. Processes for their extraction are relatively recent, but the metals are so abundant in the earth's crust that they will surely prove increasingly important in the future.

Throughout historic time, it has been the base metals, copper, tin, iron and lead and to a lesser extent mercury, zinc, nickel and antimony, in addition to gold and silver, that man has used most widely for his comforts, weapons and industry. The search for the ores of these metals, their removal from rocks, and the extraction of the metals constitute a major document in human history.

The record of the discovery and exploitation of metallic deposits is so vast and so packed with exciting stories of hardship, intrigue and adventure that it is impossible to describe in detail even the best-known stories. Iron and tin are considered elsewhere, and the following pages concentrate chiefly on minerals used as ores of the other common heavy metals.

Metal Concentrations in Nature

By means of chemical analyses of rocks found in all parts of the world, geochemists have determined the relative abundance of the ninety-two elements in the earth's crust. Some of their figures are startling. As the accompanying table shows, many of the familiar metals are present in extremely small quantities. Therefore, for a deposit to be mined economically there must be a considerable concentration. For example, even a low-grade copper ore requires at least 0.80 per cent of the metal; this means that a deposit must contain about two hundred times the average amount found in the rocks.

One of the natural processes that has been most responsible for raising the percentages of metals above the average in earth's crust is the crystallization of igneous magma. As magma cools, the most abundant elements come together to form crystals of the common rock-forming minerals. Minute amounts of the rarer metals, excluded from these minerals, concentrate in the still liquid part of the magma. In the final stages of crystallization, watery solutions enriched in minor elements are expelled through cracks and fissures in the overlying solid rock. As these solutions move toward the surface they become progressively cooler, precipitating different minerals at different temperatures. Thus the metals found in a vein may be quite different in depth from what they are near the surface. For this reason a shallow lead-zinc mine conceivably could be a copper mine at depth, and a tungsten mine at still greater depths. In addition to the ores of the metals, each depth zone contains a characteristic gangue, worthless minerals that must be removed before the metal can be extracted.

Percentage of Some Common Metals in the Earth's Crust

Silicon	27.72	Chromium	0.020
Aluminum	8.13	Zirconium	0.016
Iron	5.00	Nickel	0.0080
Calcium	3.63	Zinc	0.0065
Sodium	2.83	Copper	0.0045
Potassium	2.59	Cobalt	0.0023
Magnesium	2.09	Lead	0.0015
Titanium	0.44	Tin	0.0003
Manganese	0.10	Tungsten	0.0001
Arsenic	0.0005	Silver	0.00001
Molybdenum	0.0001	Platinum	0.0000005
Mercury	0.00005	Gold	0.0000005
Antimony	0.00002		

Concentrations of metals are related to igneous magmas in other ways. As a magma with an unusually high percentage of a metal crystallizes, minerals containing that metal may form in the resulting igneous rock. In many places such metallic minerals scattered throughout a large rock mass are present in sufficient quantity to be mined. Disseminated metals in other magmas may combine with sulfur to form tiny droplets of metallic sulfides. As these droplets coalesce and become larger, they sink toward the bottom of the magma chamber. Most of the world's nickel is recovered from deposits possibly formed in this way. Concentrations of chromium are also presumably formed in this way, but as the oxide rather than the sulfide. As the magma cools, chromium combines with iron and oxygen to form the mineral chromite which, being

Massive stibnite with the red arsenic sulfide, realgar, from Manhattan, Nevada. The single crystal of stibnite from Iyo, Japan, shows the remarkable property of bending without breaking. (Emil Javorsky)

heavier than the surrounding magma, sinks and accumulates near the bottom of the body.

The minerals deposited by hydrothermal solutions or formed on crystallization of a magma are called the primary minerals. When they are exposed at the earth's surface they react with rain water moving downward through the rocks to form secondary minerals. It is this latter type that the prospector usually encounters and during the early days of a mining operation it makes up the ore from which the metals are won. However, such surface reactions are usually shallow and the secondary minerals give way in depth to the primary ore in which all the great metal mines of the world operate.

Primary Ore Minerals: Cinnabar

Cinnabar has been known for a long time, so long that the origin of the name is lost in antiquity. Because of its vermilion-red color the mineral was used as a pigment before it was known to contain mercury; today it is known as the only ore of mercury. The physical properties of cinnabar are very distinctive; in addition to the red color it has an extremely high specific gravity.

It is deposited near the surface by hydrothermal solutions. In fact, in several places it is being deposited by hot springs, together with arsenic and antimony minerals. Mercury mines are thus shallow and none extends to any great depth. The world's oldest mercury mine, at Almadén in south central Spain, is still in operation today. It was first worked by the Carthaginians and later by the Romans. The Romans knew of the amalgamation method for recovering gold and silver, a process by which tiny particles of the metals, which otherwise would be lost, are trapped in pools of mercury. By the same process the early Spaniards in Peru used mercury from local deposits in the extraction of silver from its ores. Cinnabar deposits were discovered in California shortly after the discovery of gold and the mercury from them played an important role in mining practice following the gold rush.

Arsenic and Antimony

A mineral containing arsenic may have an antimony-bearing counterpart so closely resembling it that it is impossible to tell the two apart by simple tests. The two elements are in fact so similar in their chemical properties that they substitute for one another in a given mineral structure. Their greatest concentration is in the minerals called the sulfosalts, where they occur in chemical combination with sulfur and a metal such as silver, copper, lead or zinc. Much of the arsenic and antimony of commerce comes as a by-product from the smelting of such ores. There are, however, minerals that are mined for these elements alone.

If a single mineral were to be selected as an ore of arsenic, it would be hard, silvery-white arsenopyrite, which contains iron and sulfur as well as arsenic. In some places it occurs in large masses by itself and in others it is associated with native gold. The arsenic sulfides, realgar (AsS) and orpiment (As_2S_3), which occur in deposits near the surface, have long attracted attention because of their brilliant colors. Red realgar and lemon-yellow orpiment were used for centuries as pigmenting materials until the lethal qualities of the arsenic

(Above) Red sphalerite, called "ruby jack," from the Tri-State district, where Missouri, Kansas and Oklahoma meet. (Arthur Twomey)

(Below left) Sphalerite from Joplin, Missouri, coated with an iridescent blue tarnish and sprinkled with yellow chalcopyrite crystals. (Tad Nichols)

(Below right) Large brass-yellow crystals of chalcopyrite with sphalerite. (G. Tomsich)

in them were discovered. The use of these minerals or their synthetic counterparts as pigments is now banned in most countries. For the same reason arsenic is much less common in insecticides than formerly, giving way to compounds containing less toxic elements.

The mineral stibnite (antimony sulfide) has been known since Biblical times and was used by the ancients as a cosmetic and a medicine. Because of its low melting temperature (it fuses easily even in the flame of a match) it was a favorite material of the alchemists. Today it is of interest as the major source of antimony used in a number of alloys such as antimonial lead for storage batteries, printing type metal and pewter. Stibnite occurs in slender black crystals with lustrous faces and is distinguished from similar appearing minerals by its single perfect cleavage and the property of bending without breaking. Although the most important producing locality is in the province of Hunan, China, the choicest crystals and mineral specimens have come from a deposit on the island of Shikoku, Japan.

Galena and Sphalerite

Galena (PbS) and sphalerite (ZnS) are respectively the most important ores of lead and zinc. Both minerals form under similar conditions and are usually associated; the mine that produces lead usually produces zinc.

Galena is frequently found in well-formed cubic crystals and is easily recognized by its lead-gray color, perfect cleavage parallel to the cube faces, a high specific gravity, and a brilliant metallic luster. Galena may also contain small amounts of silver minerals and thus is often mined as much for its silver as its lead.

The mineral sphalerite derives its name from the Greek word *sphaleros* meaning treacherous, for it was known that it often resembled galena but yielded no lead. Furthermore, it is difficult to identify for, unlike galena, it occurs in many color varieties, ranging from black to colorless. American miners have given sphalerite the name "Jack" and speak of "black jack," "ruby jack" (red), and "resin jack" (yellow-brown with a resinous luster). Rarely is the mineral pure and colorless since iron is almost invariably present and darkens the mineral accordingly. In addition to iron, the less well-known metals cadmium, indium and thallium may also be present.

Deposits of galena and sphalerite have been found in nearly every country. The extraction of lead from galena is one of the simplest metallurgical processes and was known at a very early time: it need merely be thrown into a hot fire, where it melts and is reduced to the metal by the glowing coals. Pellets of lead can later be removed from the ashes. By such a method lead was obtained for bullets by the early settlers of Pennsylvania and Missouri and for water pipes by the builders of Pompeii; and undoubtedly the shapeless masses of lead found in the ruins of Hissar, now Tajik, in Asiatic Russia, were produced in a similar manner around 3000 B.C.

In contrast to lead, zinc is a young metal. The late discovery of zinc was due to the fact that the temperatures necessary to reduce the metal from sphalerite also vaporized it. Zinc was probably not produced until the end of the sixteenth century. At about the same time brass, composed of one-third zinc and two-thirds copper, made its appearance.

(Above) Sulfide minerals: (top left) chalcopyrite, a copper ore mineral, with reddish niccolite, a nickel ore, in white quartz from Ontario, Canada. (Top right) A copper mineral, bornite, from Arizona, sometimes called "peacock ore" because of its iridescent tarnish. (Bottom left) Pyrite, the most common sulfide mineral, from Butte, Montana. (Bottom right) Cinnabar, the only ore of mercury, from Arkansas. (All Ted Boente collection, except pyrite from Joseph and Helen Guetterman. Photo by John H. Gerard)

(Below) Hexagonal crystals of pyrrhotite from Kisbanya, Roumania, lie in a matrix of black sphalerite and white quartz. The copper-colored specimen is niccolite, an ore mineral of nickel from Richelsdorf, Germany. (Emil Javorsky)

Copper

At various times and places many copper minerals have served as ore minerals. The most important of these are listed below.

Important Copper Ore Minerals

	% Copper	Chemical Composition
Native Copper	100	Cu
Chalcocite[1]	80	Cu_2S
Covellite[1]	66.4	CuS
Bornite	63.3	Cu_5FeS_4
Antlerite[1]	53.8	$Cu_3(OH)_4SO_4$
Enargite	48.3	Cu_3AsS_4
Chalcopyrite	34.5	$CuFeS_2$

[1] Most commonly found as secondary minerals.

As the native metal uncombined with other elements, copper has been an important ore at only one place, the Keweenaw peninsula, Lake Superior. The other copper minerals are widely distributed; in some places only one is present, but more frequently several are found together. Each has distinctive characteristics, as follows:

Chalcocite.	Lead-gray; no cleavage; easily scratched by a knife (hardness 3), which leaves a shining furrow.
Covellite.	Indigo-blue; one perfect platy cleavage; frequently found coating other copper minerals.
Bornite.	Pinkish-colored on a fresh fracture but tarnishes purple and blue and hence is called "peacock ore."
Antlerite.	Emerald-green; good prismatic cleavage.
Enargite.	Grayish-black; good blade-like cleavage; melts in a match flame.
Chalcopyrite.	Brass-yellow; easily scratched by a knife, which leaves a greenish-black powder.

Silver

Because of its value, a mineral need contain only a small percentage of silver to be an ore. Thus, when the lead mineral galena contains small "impurities" of silver, these make it a major silver ore. Similarly, tetrahedrite, essentially a copper mineral, is sometimes sought for its silver content. However, there are many minerals in addition to native silver, especially argentite, pyrargyrite and proustite, in which silver is the major metal, if not the only one.

Argentite is black, heavy (sp.g. = 7.3) and perfectly sectile, that is, it can be cut into shavings by a knife. It contains eighty-seven per cent silver and, where abundant, is an important ore of the metal. It commonly occurs with other silver minerals, particularly pyrargyrite and proustite, known as the ruby silver ores because of their ruby-red color. These minerals form similar

(Above) Spherical masses of malachite (green) with radial structure are coated with delicate fibers of chalcotrichite, a variety of cuprite sometimes called "plush copper." (Arthur Twomey)

(Below) Needle-like crystals of millerite (nickel sulfide) form a radial group in a cavity in hematite from Antwerp, New York. (Katherine H. Jensen)

crystals and are chemically alike, but pyrargyrite, or dark ruby silver, contains antimony, whereas proustite, light ruby silver, contains arsenic.

Nickel

Three primary minerals, niccolite, millerite and pentlandite, are the major ores of nickel. The first two, found in hydrothermal veins associated with ores of silver and cobalt, are of less importance. Pentlandite, the chief nickel ore mineral, is found in dark igneous rocks associated with pyrrhotite. Although pyrrhotite itself contains no nickel, it is by far the more abundant mineral and is mined in large quantities, as at the great mines of Sudbury, in Ontario, Canada, for the small amount of pentlandite it contains. Niccolite was early mined with silver ores in Saxony, but its chemical nature was not known until 1751 when the Swedish chemist Cronstedt isolated nickel from it. Up to that time its copper-red color had led people to believe it was a copper mineral. Millerite is a comparatively rare mineral with a brass-yellow color and is frequently found in fine hairlike masses; in only a few places is it present in sufficient quantity to be mined as a nickel ore. Pentlandite has a bronze color and looks like pyrrhotite, with which it is almost invariably found. The two minerals can be easily separated on a commercial scale since pyrrhotite, known as "magnetic pyrites," is attracted to a magnet, but pentlandite is not.

Secondary Ore Minerals

The primary ore minerals which supply most of our metals and are thus of most concern to the economic geologist may be altered by natural processes

A clay-plate picture showing various aspects of mining at Corinth, Greece, about 600 B.C.

*(Right) Malachite, green copper carbonate, from the Congo. Solutions working down from the surface deposit layer upon layer of malachite in hummocky masses. If the hummocks are sliced through, the layering appears as concentric circles.
(Studio Hartmann)*

(Left) Azurite from Tsumeb, Southwest Africa, altering to malachite. The blue crystals (left) are azurite; the green crystals (right) are pseudomorphs of malachite after azurite, altered from azurite but preserving its crystal form. The crystal of azurite (center) is partially altered to fibrous radiating malachite. (Emil Javorsky)

(Below far left) Rust-colored crystals of quartz and colorless crystals of calcite from Mexico are partially concealed by a coating of green malachite. (Studio Hartmann)

(Below left) The blue of chrysocolla from Globe, Arizona, is set off by brilliant stains of iron. (Katherine H. Jensen)

(Right) Cubo-octahedral galena crystals growing on a large cubic galena crystal from Treece, Kansas. (Elmer B. Rowley)

to form "secondary" or supergene minerals. These secondary minerals have more appeal to the mineralogist and mineral collector partly because of the interesting geochemical processes that produced them, but mostly because they are found in well-formed crystals, often with brilliant colors.

As rainwater, with dissolved oxygen and carbon dioxide, slowly percolates down into rocks from the surface, it reacts with the primary ores and enclosing rock to produce new compounds. Some of these are carried away in solution and eventually reach the sea, others may be deposited deeper down, in the rocks, and still others may remain in place as new minerals. These secondary minerals near the surface indicate the metals one may reasonably expect to find in primary minerals at greater depth.

The formation of these secondary minerals is promoted by the presence of the iron sulfide, pyrite; although not an ore itself, pyrite is universally associated with the sulfide ore minerals in so-called vein deposits. In contact with surface waters, it yields iron sulfate and sulfuric acid that greatly accelerate the solution of other minerals. In addition, its alteration leaves behind the hydrated iron oxide, limonite, a rusty-colored mineral that is the principal and perhaps the only surface indication that valuable ore minerals may be present in depth.

Copper is present in the largest, most varied and colorful group of secondary minerals. The copper oxide, cuprite (Cu_2O) forms in cubic crystals of a ruby-red color. It has a brilliant luster and were it not for the low hardness ($3\frac{1}{2}-4$) its transparent crystals would make lovely gems. Sometimes associated with cuprite are the copper carbonates, malachite and azurite, whose respective bright green and intense azure-blue make these the most spectacular of all secondary minerals. They are often found together, but malachite is more

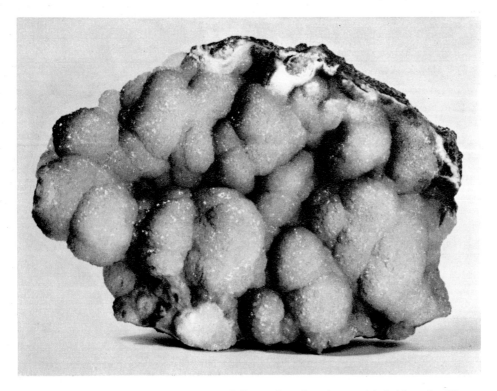

Brilliant-faced hemimorphite crystals, the zinc silicate, coat the irregular surface of a cavity in a specimen from the oxidized zone of the zinc mine at Ogdensburg, New Jersey. (Elmer B. Rowley)

common. It occurs in aggregates of fibers forming botryoidal (that is, like a bunch of grapes) or stalactitic masses. Although too soft for jewelry, it has for centuries been cut and polished for decorative purposes. Small objects such as beads or trays are made from massive malachite, but more commonly it is cut into thin slabs to be used as an inlay or veneer. The outstanding locality for this ornamental material is near Nizhni Tagil in the Ural Mountains. One may see in the Hermitage Museum in Leningrad the artistry with which early Russian craftsmen veneered exquisite urns and table tops with matched pieces of polished malachite. Small amounts of malachite are found at many places but only a few have furnished it in quantity for commerce or in specimens worthy of museum exhibit. The most important sources are Tsumeb in Southwest Africa, Katanga in the Congo, the Ural Mountains in Russia, and Bisbee, Arizona.

Azurite and malachite are so alike chemically that a small variation in the surrounding moisture is sufficient to make one change to the other. Thus if azurite is found in botryoidal masses, it may be assumed that it has been altered from malachite. Unlike malachite, azurite often occurs in crystals of characteristic shape, some as large as three to four inches across. When, therefore, one sees malachite in the crystal form of azurite, it is a pseudomorph, that is, azurite altered to malachite. At one time, finely ground azurite was used by artists as a paint pigment but it was most unsatisfactory, for the blues of the painting would frequently turn green after a few years as azurite altered to malachite.

One of the most important copper ore minerals, chalcocite, may form either as a primary or secondary mineral but its chief commercial deposits are of secondary origin. Copper dissolved from minerals near the surface is carried downward and precipitated as chalcocite. The lower limit of precipitation is

(Right above) A black core of galena from Tiger, Arizona, is surrounded by its alterations products, wulfenite (yellow-orange) and cerussite (gray). (Emil Javorsky)

(Below) Smithsonite, usually white, from Laurium, Greece, is colored blue by a small amount of copper. (Katherine H. Jensen)

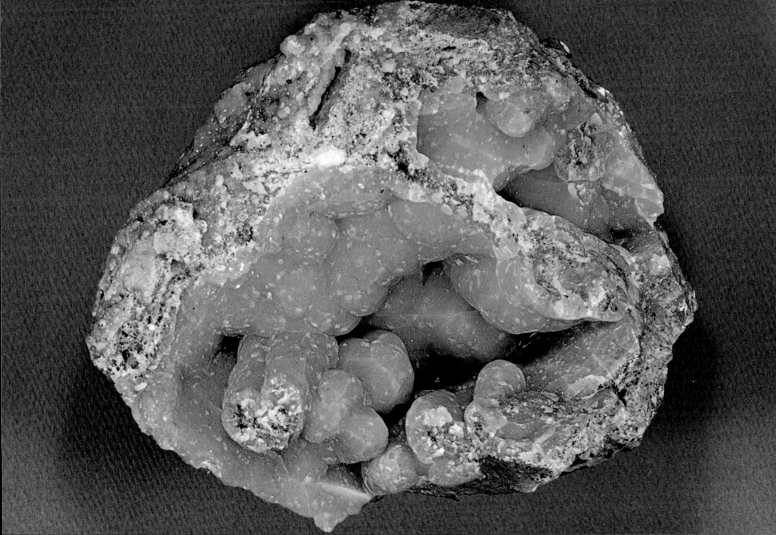

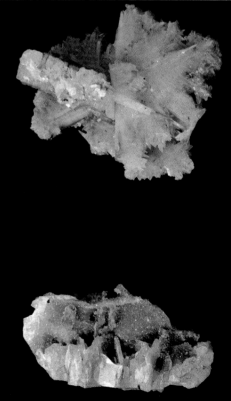

(Left) Blue needles of aurichalcite and white crystals of calcite on a background of yellow smithsonite. From Pinal County, Arizona. (Reo N. Pickens, Jr.)

(Below far left) Orange-red crystals of vanadinite from Globe, Arizona. (Floyd R. Getsinger)

(Below left) Cerussite from Tsumeb, Southwest Africa, in a group of reticulated crystals (above) and encrusted with green smithsonite (below). (Emil Javorsky)

the water table; below which the rocks are permanently saturated with water. This irregular surface separates the zone of primary minerals below from the zone of secondary minerals above. In many places a rich "blanket" of chalcocite has formed at or near the water table and has produced a rich copper deposit. At other places, this process has precipitated a small amount of chalcocite on scattered grains of primary minerals, making a worthless deposit into one that is worth mining.

The secondary copper minerals formed in arid regions differ from those found in more humid parts of the world. The former are largely copper sulfates, soluble in surface water and therefore formed only in areas of low rainfall. Most of these sulfates such as blue vitriol, the mineral chalcanthite, are rare. However, antlerite, a secondary copper sulfate, is the chief ore mineral of the world's largest copper mine. This great mining operation is located at Chuquicamata, Chile, in the Atacama desert, one of the most arid places in the world.

Lead and zinc, universally and intimately associated in the primary sulfide minerals, galena and sphalerite, are separated by the action of surface water. Zinc sulfate, formed from sphalerite, is soluble and thus is carried away in solution. Lead sulfate, formed from galena, is insoluble and may remain in place as the mineral anglesite; or cerussite (lead carbonate) may result if the ores are in limestone. More rarely pyromorphite (lead phosphate), mimetite (lead arsenate), vanadinite (lead vanadate), and wulfenite (lead molybdate) are present as secondary minerals in the oxidized portion of lead veins. These minerals have a high luster and frequently occur in well-formed crystals but only occasionally in sufficient quantity to be ores of lead. They make superb exhibit specimens and are coveted by mineral collectors. This is particularly true of vanadinite and wulfenite with their rich red, orange and yellow colors.

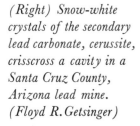

(Right) Snow-white crystals of the secondary lead carbonate, cerussite, crisscross a cavity in a Santa Cruz County, Arizona lead mine. (Floyd R. Getsinger)

The two supergene zinc minerals of most interest are hemimorphite, the silicate, and smithsonite, the carbonate. If solutions containing zinc sulfate, produced by oxidation of sphalerite, pass through silicate rocks, hemimorphite may form. It frequently occurs in crystalline aggregates forming crusts that sparkle as light is reflected from the many crystal faces. Smithsonite is a more common and abundant mineral. If the primary minerals are found in limestone, as they commonly are, the solutions containing zinc react with the limestone to form smithsonite. In some areas these secondary deposits are extensive and are valuable zinc ores. In general, smithsonite is a dirty brown and, resembling the surrounding rock, may pass unnoticed by the prospector. However, in rare instances it is a translucent bluish-green and has been used in jewelry and for ornamental purposes. This decorative material has come from Kelly, New Mexico; Tsumeb, Southwest Africa, and Laurium, Greece. A yellow variety of smithsonite, containing cadmium, is called turkey-fat ore; it has been found in Sardinia in stalactites.

The ore in the great copper mine at Bingham Canyon, Utah, contains less than one per cent copper but so vast a tonnage is mined that Bingham Canyon is the world's largest copper producer. (John S. Shelton)

Secondary silver minerals are much less common than those of copper, lead and zinc. Even when rich silver ores are present in depth, there is little surface evidence of silver. The most important secondary silver mineral is cerargyrite (silver chloride). In a few places large amounts of cerargyrite associated with smaller amounts of bromyrite (silver bromide) and iodyrite (silver iodide) are found. These minerals all have similar characteristics, and are colorless when pure and fresh but on exposure to light become a violet-brown or purple. They look like the horn of an animal, and like horn, can be easily cut with a knife and thus have been called "horn silver." In some places native silver is found with cerargyrite.

Although the metals used earliest by man were those that occurred in the native state, that is, uncombined with other elements, some early mined ores required a metallurgical process to extract the metal. After the native metals, the secondary minerals were used; for they not only occurred near the surface but the metallurgy required was simple. To win the metal from many supergene copper, lead and silver minerals, the ore need only be placed in a hot fire of glowing coals to remove unwanted elements and produce a molten metal.

To recover metals from most primary minerals or to separate one metal from another requires a more elaborate and sophisticated process. Such methods must have been known to the Greeks in 1000 B.C. and perhaps to the Phoenicians a thousand years earlier. From mines at Laurium, in Attica, came a wealth of silver on which, at least in part, rested the greatness of ancient Athens. The ore was essentially silver-bearing galena but contained some sphalerite. The minerals were separated from one another and the sphalerite discarded with the worthless rock. In smelting the galena, the silver was separated from the lead. Some of the slag and waste material dumped into the sea has been re-excavated in recent times. The chemical reactions resulting from more than two thousand years of immersion in sea water produced many interesting minerals; laurionite, named for the locality, was found in this association.

Recovering Minerals From Low-Grade Ore

Not only early mining but nearly all mining until the beginning of the twentieth century was confined to high-grade ore bodies. Veins were followed until they terminated in depth or the proportion of metal dropped below a minimum for profitable operation. The great mining districts of the world have acquired their reputation from the high-grade ore mined in them. A few of the outstanding ones are the silver mines of Freiberg, Germany; Guanajuato, Mexico; Cobalt, Ontario, Canada and Virginia City, Nevada; the zinc-lead-silver mines of Idaho and Colorado; Broken Hill, Australia and Tsumeb, Southwest Africa; the copper mines of Butte, Montana; the Copper Belt in Zambia and Cerro de Pasco, Peru. Many of these mines are still in operation, producing not only one or two metals but several, for the ore minerals tend to occur together.

For a long time, experts knew of large masses of rock that contained metallic minerals, particularly copper, but with metal content too low to warrant mining by the methods used for high-grade ores. Today low-grade ore is mined economically on a large scale because of "flotation," a method of separating valuable minerals from waste rock. In this process a watery mixture of finely ground rock is run into tanks filled with water. Small amounts of reagents are added, depending on the mineral sought; when the water is agitated, these cause air bubbles to cling to certain minerals but not to others. The mineral particles buoyed up by the air rise to the top and are removed; the others, usually waste material, sink.

Much of the world's copper is now recovered from low-grade ore, or, as they are called, "porphyry copper ores." Over half of the copper production of the United States, the world's largest copper producer, comes from low-grade "porphyry copper ores." In some districts, where for decades mining has been selective, it is found that rock between the high-grade veins contains sufficient copper to permit large-scale mining and recovery of a small percentage of metal by flotation.

Similarly, molybdenite, the ore of molybdenum, a soft, scaly mineral much resembling graphite, is disseminated through huge volumes of fractured granitic rock and can also be recovered economically by means of flotation. Such is the type of the world's major deposit of molybdenite, high in the Rocky Mountains at Climax, Colorado.

Aluminum

Unlike most other metals, which occur sparingly in the earth's crust, aluminum is plentiful. It is a major constituent of feldspar, the world's most abundant mineral, and of the clays formed by the alteration of feldspar. In addition, it is present in untold quantities in sedimentary rocks in the vast accumulations of shale. But at the moment the only economic source of aluminum is bauxite.

Bauxite is, properly speaking, a rock and, like most rocks, is made up of several minerals. Until recently, bauxite was considered a mineral with a definite chemical composition, but it is now known to be a fine-grained aggregate of minerals, including varying proportions of the hydrous aluminum oxides—gibbsite, boehmite and diaspore. Bauxite is a residual deposit resulting from the weathering of high-alumina rocks in a tropical or semi-tropical climate. In temperate zones the chief weathering product of such rocks is clay, but in the tropics most of the constituents of the rock, such as silica, soda, lime, potash and magnesia, are removed, leaving only alumina and iron. Thus in regions where the underlying rock is high in alumina, vast areas are covered with a blanket from one to sixty feet thick of an aluminum-rich concentrate, bauxite, with iron as the only major impurity.

In the method developed late in the nineteenth century for extracting aluminum, the bauxite is dissolved in a bath of molten cryolite and the aluminum precipitated electrolytically. Since large quantities of electricity are essential to this process, the ore may be shipped thousands of miles to a source of cheap power. Thus, Canada, with no bauxite of its own but with abundant hydroelectric power, is a major producer of aluminum using ore mined in Surinam and neighboring countries in northern South America, the world's chief suppliers.

(Above) Hemimorphite crystals on limonite from Mapimi, Durango, Mexico.
(Katherine H. Jensen)

(Below left) Bauxite, from Arkansas, is composed of brown concretionary masses embedded in a gray earthy matrix.
(Benjamin M. Shaub)

(Below right) Tabular crystals of wulfenite from Santa Cruz County, Arizona.
(Floyd R. Getsinger)

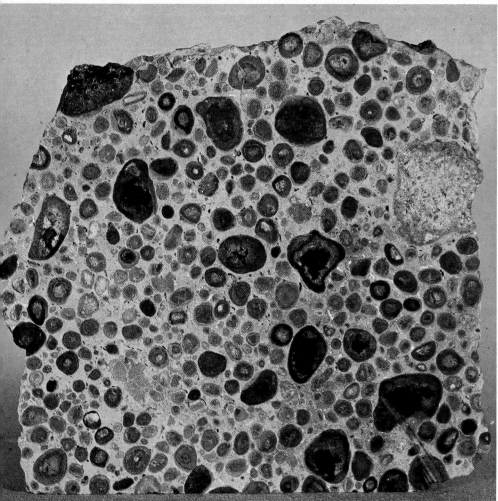

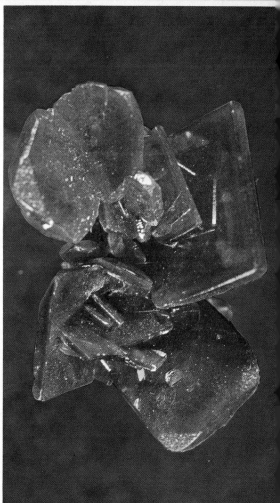

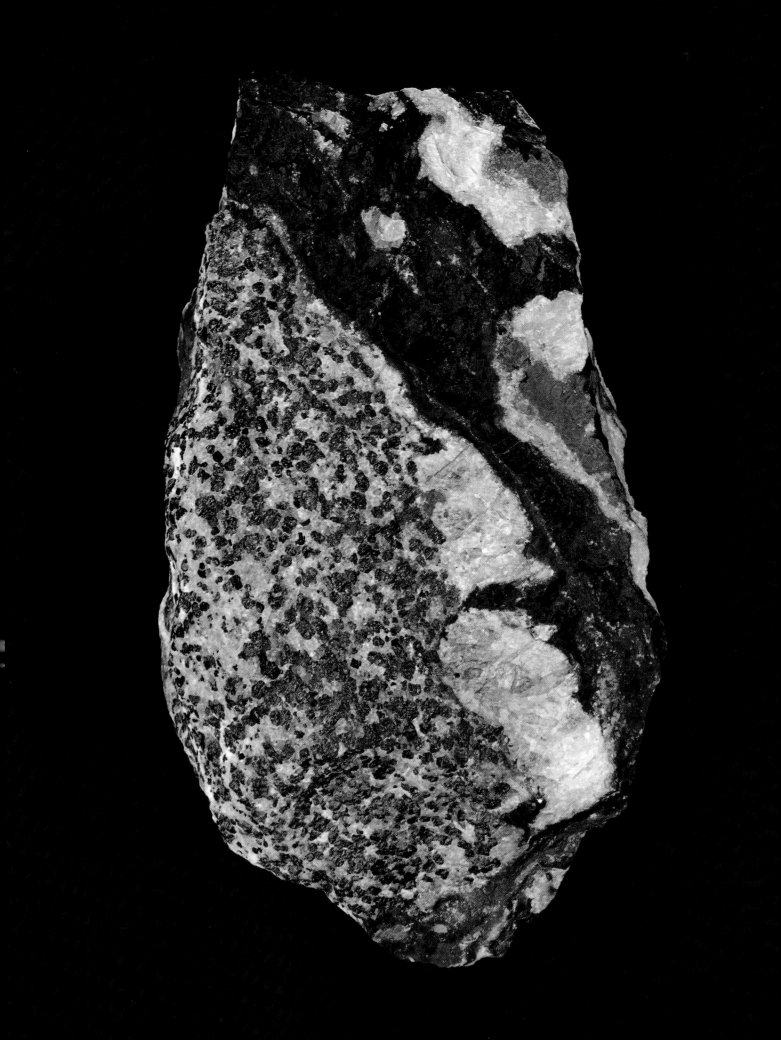

12 Minerals That Glow in the Dark

One of nature's perversities has been the location of mineral deposits in inaccessible and forbidding places. Many major mines of the world are located in—the desert, the tropics, the arctic or high mountains—areas where living and working are difficult for man. An outstanding exception is Franklin, New Jersey. Here, in the wooded hills of northern New Jersey fifty miles from New York City, is one of the great zinc deposits of the world.

There is a more striking difference between Franklin and the other major zinc deposits: sphalerite (zinc sulfide), elsewhere the universal primary ore of zinc, is almost totally absent. In its place there are three zinc-bearing minerals which are, in order of abundance, franklinite, willemite and zincite. Not only is this association peculiar to Franklin, but of the three minerals only willemite is found elsewhere. These ore minerals must be separated from one another as well as from the white calcite in which they are embedded before they are smelted.

Franklinite, an oxide structurally similar to magnetite, contains iron and manganese in addition to zinc. Physically it resembles magnetite in that it is black, crystallizes in octahedrons, and is magnetic, but only slightly so. Willemite (zinc silicate, Zn_2SiO_4) usually occurs in granular aggregates in a variety of colors. Transparent crystals of it are colorless, green or yellow; nontransparent grains are white, green, yellow, red, brown or even black. A common characteristic of nearly all willemite, despite its great range of color, is that it fluoresces a brilliant green under ultraviolet light. Zincite (zinc oxide, ZnO) as a pure chemical compound is white or colorless, but at Franklin a small amount of manganese and structural imperfection in the mineral give it a deep red to orange-yellow color.

Although the minerals of Franklin were not scientifically investigated until

Franklinite, willemite and zincite are the principal ore minerals of the great zinc mine at Franklin, New Jersey. A typical association shows black franklinite, brown willemite and red zincite in a matrix of white calcite. (Emil Javorsky)

early in the nineteenth century, it was probably one of the first ore deposits to attract attention in North America. As early as 1640, the Dutch were mining copper in the Delaware River area. Like others who followed them, they probably mistook zincite, the red zinc oxide, for copper ore. In 1772 Lord Stirling, who had acquired the pits from which the ore was dug, shipped several tons of it to England to be smelted for its copper content. The ore was apparently never smelted but specimens from it found their way into European mineral collections with the origin erroneously credited to various European localities. Since magnetite as a major ore of iron was being extensively mined in New York, New Jersey and Pennsylvania, it is not surprising that franklinite, resembling magnetite in most of its properties, was confused with it. Lord Stirling mined and shipped a large amount of franklinite to iron smelters but the zinc and manganese in it rendered it useless as an iron ore. An iron furnace built at Franklin met with failure for the same reason.

The first step in understanding the minerals at Franklin was taken in 1810 when Dr. Archibald Bruce of New York City correctly described zincite. The more abundant franklinite he assumed to be magnetite, and it remained for the French chemist Berthier, in 1819, to discover its true nature and name it franklinite, after the locality. The first correct description of willemite, the third major ore mineral at Franklin, was made in 1824 by Vanuxem and Keating under the name siliceous oxide of zinc. But the process of separating the metals from the ore still remained an enigma.

The Franklin mining district contains two main ore bodies with similar mineralogy and nearly identical form. One was located at Franklin (Mine Hill) and the other 2½ miles to the southwest at Sterling Hill. During the first half of the nineteenth century the mineral rights to these deposits were largely controlled by Dr. Samuel Fowler. He spent most of his life trying to find ways of exploiting them, but he met with little success and the mining properties passed to his son, Colonel Samuel Fowler. Although many years had passed since the mineralogical nature of the ores had been determined, early misconceptions still lingered on, and when Colonel Fowler in 1848 entered into an agreement with the Sussex Zinc and Copper Mining and Manufacturing Company, he conveyed "all the zinc, lead, copper, silver, and gold ores, and also all other metals or ores containing metals, except the metal Franklinite, and iron ores when it exists separate from zinc." Two years later he conveyed to another company, the Franklinite Mining Company, "All the reserve iron ore called Franklinite and all the reserve ores and minerals not granted . . . earlier." It was soon recognized that franklinite contained zinc, and since Colonel Fowler had granted the right to mine franklinite to one company and the right to mine zinc to the other, both companies claimed the entire deposit. For nearly forty years thereafter the ownership of the mines was in litigation. It was not until 1897 that all mining interests were consolidated in one company and the ore deposits began to be exploited efficiently. Although the ore body at Franklin was worked out by 1957, mining still goes on at Sterling Hill and promises to do so for many years.

The Franklin district is noted as a major producer of zinc from a unique assemblage of ore minerals. But to the mineralogist it is, in addition, a locality unsurpassed for the number and variety of its minerals. The geological processes that gave rise to such unusual ore minerals also produced a host of other

White calcite and brownish willemite (below) from Franklin, New Jersey, emit brilliant colors when exposed to ultraviolet light (above), the calcite fluorescing a brilliant red and the willemite a vivid yellow-green. (Both by Katherine H. Jensen)

rare minerals. The many collections of Franklin minerals made during the past century and a half are a continuing source of new discoveries, including minerals completely new to science. Over 160 minerals have been reported from the Franklin district. Even more remarkable is the fact that forty of them were first described from there and that only three or four of these have since been found at other localities. Many of these unusual minerals are present only in small amounts, but it should be emphasized that franklinite and zincite, occurring at Franklin in thousands of tons, have never been found elsewhere.

At most mines the ore broken underground goes directly to a crusher and then to a mill for mineral separation. At Franklin, however, the ore was dumped onto a "picking table" before going to the crusher. The table was slowly rotated while several men picked off and discarded waste material and minerals containing the objectionable elements, lead and arsenic. At the same time they became skillful in identifying unusual minerals; many a collection is rich in rare Franklin minerals because of the keen eye of one of these self-trained mineralogists.

Since it was impossible to mine franklinite, willemite or zincite separately, it was early realized that a method had to be devised for separating the minerals so that each could be treated by itself.

John Price Wetherell, one of the mine owners, became interested in this problem and in 1888 devised a magnetic separator for treating the ore at Mine Hill. The franklinite, which is slightly magnetic, is attracted to the poles of a strong electromagnet, leaving behind the nonmagnetic zincite, willemite and calcite. Today the Wetherell magnetic separator is used for ore dressing in all parts of the world. The zincite and willemite are then separated from the calcite and other waste material on "shaking tables" which sort out minerals according to their specific gravity.

Fluorescent Minerals

Fluorescence is the capacity of some substances to emit visible light when excited by short-wave radiations—a phenomenon made familiar by fluorescent lighting tubes. The most common means of excitation is ultraviolet radiation; since these rays are invisible to the human eye, their effect on a strongly fluorescent mineral is most dramatic. When the invisible rays are turned on, the specimen glows in the dark. When the radiation is turned off, the fluorescence in most minerals stops immediately, but some continue to glow for several seconds or even minutes. This lingering luminescence is called phosphorescence. Although fluorescence from exposure of minerals to sunlight was reported early in the seventeenth century, it was not until 1852 that Sir George Stokes, who experimented with fluorite under the sun's rays, explained the phenomenon scientifically. It was he who gave the name of the mineral to this interesting property.

Many minerals fluoresce but none more brilliantly or with a greater variety of color than those from Franklin, New Jersey. The discovery of fluorescence in Franklin minerals took place early in the twentieth century, quite by accident. A workman near the shaking tables where the mineral grains were being separated, pulled a switch to shut off the electric power to the tables. The workman noticed that each time the switch was pulled a pale blue arc

Brown cubic crystals of fluorite and white-to-colorless blades of celestite (below) undergo a dramatic color change when viewed in ultra-violet light (above). The fluorite gives off an eerie green light and the celestite a vivid blue (Both by Reo N. Pickens, Jr.)

was generated causing bright yellowish-green spots to appear on the tables. On further examination it was found that the grains glowing in the dark were willemite. It was soon realized that this property could be used to distinguish the willemite from the waste calcite. An instrument called an iron-arc spark-gap was constructed that generated a feeble ultraviolet light. Until replaced by more efficient sources of ultraviolet light, the iron-arc was used to determine the presence of willemite in mill-tailings. It was also used as late as 1930 to test for fluorescence in other minerals.

It is impossible to predict whether a mineral will fluoresce when subjected to ultraviolet radiation. But the phenomenon appears to be related to minute amounts of impurity in the minerals.

At Franklin the chief impurity is manganese. When pure, willemite contains only zinc, silicon and oxygen; it is quite colorless and does not fluoresce. With only trace amounts of manganese, the mineral glows a vivid yellow-green, under ultraviolet light, and will continue to phosphoresce for several minutes after the radiation has been removed.

Pure calcite does not fluoresce but at Franklin it contains a small amount of manganese, resulting in a striking red fluorescence like glowing coals. Calcite from Långban, Sweden, also contains small amounts of manganese and fluoresces a similar color. Among the less common Franklin minerals, the rare mineral calcium-larsenite gives off a vivid lemon-yellow; pectolite fluoresces a pale-yellow; and clinohedrite an orange-yellow.

The fluorescent property of minerals has other applications. Thus a prospector using a portable ultraviolet light at night can quickly detect the mineral scheelite, a valuable ore of tungsten, which has a constant pale-blue fluorescence but in daylight resembles the quartz with which it may be associated. A miner underground can also use fluorescence on a freshly blasted surface to determine whether a mineral is present in sufficient amount to warrant further mining. The lithium aluminum silicate, eucryptite, is a rare mineral except at Bikita, Rhodesia, where it occurs in the hundreds of tons. Its physical properties resemble those of quartz so closely that it is impossible to tell the two apart by inspection. However, under ultraviolet light it fluoresces a vivid salmon-pink and stands out from other minerals. At Bikita the ore is passed under ultraviolet lamps and the eucryptite easily picked out. Many but not all diamonds fluoresce a pale-to deep-blue color and many uranium minerals have a characteristic fluorescence that aids in their determination.

The systematic study of fluorescence in minerals, begun at Franklin, has become so important that if a mineral has a fluorescent color, this property must be included in its complete mineralogical description. As a result of routine examination in ultraviolet light, many minerals that formerly passed unnoticed have been detected and described as new species. Solely from an aesthetic point of view, a well-chosen group of minerals makes a delightful and often stunning display, since drab lifeless stones suddenly became luminous and burst into a variety of brilliant colors.

13 Iron and Its Ores

In Biblical times, according to the Book of Exodus, "the Lord went before them by day in a pillar of cloud, to lead them the way; and by night in a pillar of fire, to give them light." In our time, the fiery stacks of the blast furnaces by night and the open-hearths painting the sky red by day lead the way in our twentieth century industrial civilization, for we live in the age of iron. It may be that in time to come another metal will become dominant, but today molten iron and steel are the life blood in the veins of our civilization.

The complexity of our modern world requires, and in part results from, the use of metals little known at the beginning of the twentieth century, such as titanium, uranium, vanadium, molybdenum, chromium and beryllium. But the fact that the blast furnaces in the United States alone pour forth more than one hundred million tons of iron a year indicates its overpowering importance. Most of this staggering quantity of metal will be adjusted in composition and heat-treated to make steel, of which the framework of our civilization is literally built. Practically every large country in the world will also produce vast quantities of steel (though none as much as the United States), and possession of ores of iron or their easy access will determine their industrial strength and vitality.

Man and Iron

Although iron can be released from its ores at temperatures lower than those necessary for smelting copper, the smelting of iron was not practiced to any appreciable extent until at least a thousand years after the recovery of copper from its ores had become well-established. Hematite, the most widespread iron ore, was used in very ancient times for making beads and cylinder seals, and the

Nature has stacked glistening hexagonal plates of hematite one upon the other to build up this composite crystal from Val Tavetsch, Switzerland. (Reo N. Pickens, Jr.)

pulverulent red variety was used as a pigment. The development of an iron technology was thus long delayed not by the lack of ore but probably by the nature of the smelting process itself.

Iron can be reduced from hematite at temperatures well below 1000° C (1830° F), but at this temperature the iron is not liquid, but forms with admixed slag and dirt a porous, pasty gray mass, scarcely suggesting the presence of metal at all. However, if this mass is kept red-hot and hammered vigorously, the small droplets of iron will weld together and eventually furnish a lump or bar of reasonably useful wrought iron. This discovery was probably made before 1400 B.C. somewhere in Anatolia, the land of the Hittites. Sporadic occurrences of man-made iron of earlier date may represent earlier discoveries that were forgotten and that seem to have no link with later developments.

Buried with the boyking Tutankhamen in 1350 B.C. were a dagger with a blade of iron, a set of tiny miniature iron tools and a miniature iron head rest, all crudely made of welded pieces of man-made iron. The placing of iron objects among the treasures of a king whose priests could afford to place his body in a coffin of solid gold weighing 242 pounds indicates the high scarcity value and prestige enjoyed by the rather drab-looking new metal. By 1200 B.C., iron seems to have been fairly widely known. But it was still regarded as a precious metal, and its use was restricted to jewelry, and ceremonial and ritual objects. In spite of this scarcity and the technical difficulties of smelting, ironmaking is far less difficult and a less specialized task than bronze smithing.

The Jur, a tribe of the southern Sudan, make iron today in small furnaces shaped like egg cups about the size of a man, using charcoal and hematite without forced draft. They improve their product by a second melting, using still smaller furnaces and forced draft from primitive bellows. The method is probably, like the Jur smiths themselves, directly descended from ancient Egyptian times. There is thus no doubt about the early methods of smelting iron, nor about the success of the Jur smiths in obtaining the metal. The difficulty lies in the treatment of the "bloom" or mass of iron that results from smelting, reheating and hammering. Untempered and untreated, iron is a less satisfactory metal than good bronze. Adjustment of the carbon content, proper quenching, and skillful tempering by reheating are necessary to obtain a product that is neither too soft to be used in cutting, nor so hard that it will shatter. The early blacksmith's occasional successes were attributed to supernatural intervention, and the occasional good weapon became a legend, like Excalibur, the sword of King Arthur. The Jur smiths to this day surround their ironmaking with elaborate ceremonials, but nevertheless produce an iron that is too soft to be practical for everyday tools and hence is used chiefly as a medium of exchange.

Throughout ancient Roman times, iron smelting was conducted on a small scale, and iron never wholly supplanted bronze in the esteem of Romans. A "bloom" of Roman date weighing 344 pounds was found in England near Hadrian's Wall. Such a mass was built up by welding and hammering together many smaller pieces, and might well have represented a full day's output of one of the larger Roman smelters. Iron came to Britain late, a thousand years after it was current in the Middle East, and remained a rare metal in many parts of Europe until the Middle Ages. Typical of northern European and British ironwork are the hundreds of fearsome steel swords, spear points and axe heads recovered from weapon hoards such as that found at Nydam Moss in Denmark. The long iron swords found in the cemeteries of the Hallstatt culture in Austria and Bavaria are famous, but the quality of the metal was not uniform and was generally poor.

Iron technology in the Eastern world was as far advanced as that of the West. In India, iron was made in considerable quantities before the advent of the Christian era. Two wrought-iron pillars, one twenty-three and the other forty-two feet high, erected about A.D. 310, are still objects of wonder today. They were apparently made by forging and welding discs from twelve to sixteen inches in diameter into a solid column.

In China, iron was made from early times, and the art of casting iron by addition of phosphorus to increase its fluidity was mastered there at a relatively early date. What is alleged to be the largest cast-iron statue ever made was poured in China about A.D. 950 and is eighteen feet long and twenty feet high.

In the numerous detailed, charming and informative woodcuts that illustrate the works of Georgius Agricola (of these works the *De Re Metallica,* published in 1546 and translated into English by the later President Herbert Hoover and Mrs. Hoover, is best known) we find a complete portrayal of the mining and smelting practices of the sixteenth century. These differed little from earlier Roman practice and underwent little change until the middle of the nineteenth century. The restorations of two early iron smelters and foundries in the United States give a picture of the iron industry in the seventeenth and eighteenth centuries that differs little from the one pictured in the pages of *De Re Metallica.* At Saugus, Massachusetts, an ironworks that operated between the years of 1646 and 1670 has been completely restored, including a blast furnace, rolling mill, slitting mill, forge and other buildings. The ore used in this earliest successful ironmaking venture in the New World was "bog ore," the mineral limonite. The finding of such iron oxide ore as bottom deposits in the swamps of the Saugus River prompted John Winthrop, Jr., son of the first governor of the Massachusetts Bay Colony, to take samples back to England and solicit investments in the "Company of Undertakers for Iron Works in New England." He was successful, and with the Saugus works, the American iron and steel industry was born.

Beneath the rolling hills and fertile fields of Pennsylvania lie rich deposits of iron ores that crop out near Lancaster, Pennsylvania. They were mined in shallow pits in the early eighteenth century, and smelted at nearby Hopewell Furnace, one of hundreds of small ironworks. Hopewell Furnace has been fully restored as a National Historic Site and presents a complete picture of late eighteenth and early nineteenth century ironmaking. The ore is chiefly magnetite, occurring as streaks and masses in metamorphosed sedimentary rocks, and was mined within a few miles of the furnace. Until the advent of the Bessemer process for steelmaking in 1856, practically all the iron used in the United States—perhaps five hundred tons a year—was produced by such small self-sufficient communities as Hopewell Furnace.

At the same time as the Bessemer process came the discovery in the Lake Superior region of vast deposits of high-grade hematite ores, and by 1870 Hopewell Furnace, like most of the other small ironworks east of the Appalachians, put out its fires for good.

The Lake Superior ores of Minnesota and Michigan were originally ancient sedimentary deposits of pre-Cambrian seas later leached of their dominant silica so that almost pure iron oxide, hematite, was left. These deposits were at first mined by conventional underground methods, but in 1888 Cassius Merritt found the first chunk of high-grade ore on Mis-Sa-Be Heights, and by 1892 the Merritt family had gained ownership of the Mesabi Range and was preparing to open the Mountain Iron and Biwabik mines. The flat-lying, bedded character and shallow depth of the deposits, overlain only by a layer of unconsolidated glacial debris, seemed an obstacle to their Cornish miners, trained in hard-rock vein mining. The miners told them: "You can't mine it—no hangin' wall, no foot wall—wheer's yer granite, yer grinstun? This stuff may be ore, but yer mine can't be a mine—that's all theer is about it." Then the Merritts had an inspiration that changed the face of American mining: they would mine it with steam shovels, in an open pit! The Mesabi Range was mined this way for the next seventy years, and today perhaps eighty per cent of all metal mining is carried out by open-pit, mass-mining methods, and the percentage is increasing.

The Minerals of Iron

Among the elements of the earth's crust, iron stands in fourth place in order of abundance, making up about five per cent of the whole. Most of this tremendous amount is locked in half a dozen rock-forming minerals scattered through millions of cubic miles of rock and is quite inaccessible. Of the nearly three hundred iron-bearing minerals, only hematite, magnetite, goethite, and siderite contain sufficiently high percentages of iron and are abundant enough to be considered major ores today. Some of the properties of these and several other important iron minerals are listed in the following table:

The Hull-Rust-Sellers mine (photographed in 1951) on the Mesabi Range of Minnesota was for many years the world's largest producer of iron ore. The ore mineral, hematite, gives its red color to the walls and floor of this man-made canyon. (Cornelius S. Hurlbut, Jr.)

Iron Ore Minerals

Mineral	Chemical Composition	% Iron	Color	Hardness	Specific Gravity
Native iron	Fe	100	Gray	4½	7.6
Magnetite	Fe_3O_4	72.4	Black	6	5.18
Hematite	Fe_2O_3	70.0	Red, black	6	5.26
Goethite	$HFeO_2$	62.9	Brown	5½	4.37
Pyrrhotite	$Fe_{1-x}S$	60.±	Bronze	4	4.6
Siderite	$FeCO_3$	48.2	Brown	4	3.85
Pyrite	FeS_2	46.6	Yellow	6	5.02
Marcasite	FeS_2	46.6	Lt. yellow	6	4.89
Ilmenite	$FeTiO_2$	36.8	Black	6	4.7

Native Iron

Two kinds of natural metallic iron are found at or near the earth's surface: terrestrial and celestial. The terrestrial forms as a mineral in the rocks whereas the celestial falls to the earth's surface from outer space. Because of its great affinity for other elements, particularly oxygen, iron easily forms compounds, giving us many iron minerals. Only unusual circumstances liberate it from its combining elements to form native terrestrial iron. Such iron is thus a great rarity and is found in only a few localities where iron-rich lavas have over-ridden forests and the assimilated carbonaceous material has reduced the iron. Of much wider distribution, but no less a rarity, are the iron-rich meteorites that occasionally fall to the earth's surface. These visitors from outer space, made into tools and weapons by primitive man, can be considered the first ores of iron.

In the Great Pyramid at Giza, in Egypt, archeologists have found iron objects that date from the time of construction, five thousand years ago. Some archeologists have interpreted these and other similar finds as indicating an early iron industry in Egypt; but any of these ancient tools and ornaments that have been tested have proved to be made from celestial iron. All meteoric iron contains from five to twenty-five per cent nickel (and is more properly called nickel-iron) whereas that extracted from ores contains no nickel. If a simple test establishes the presence of nickel, this is sufficient proof that the iron is of celestial origin.

In addition to the use of meteoric iron in Egypt, there is abundant evidence in Argentina, South Africa, Greenland and India that early man beat it into weapons and utensils. Small amounts of it were used throughout historic time but the evidence rests largely on the artifacts themselves, for the written record is surprisingly scanty. The first reference, in the *Philosophical Magazine* in 1803, describes two sabers, a knife and a dagger made in India from meteoric iron that fell in 1620.

The world's most celebrated find of meteoric iron is in the United States at Canyon Diablo (Meteor Crater or Barringer Crater) in Arizona. Here, formed in the flat-lying sandstones of the Coconimo Plateau is a giant pit 4150 feet from rim to rim and 470 feet deep. Within a radius of five miles of this crater men have picked up forty thousand fragments of nickel-iron weighing more

Radiating crystals of goethite, iron oxide, from an iron mine at Negaunee, Michigan.
(Emil Javorsky)

than twenty tons, not counting the iron that was undoubtedly removed before records were kept. The evidence is that Meteor Crater was made by the impact of a gigantic iron-nickel meteorite of which the recovered fragments represent but an insignificant fraction. Most scientists today believe that the bulk of the iron was vaporized by the high temperature generated on impact. However, from 1902 until his death in 1929, D. M. Barringer, who obtained possession of the crater, believed that somewhere beneath the rim of Meteor Crater lay millions of tons of nickel-iron with some platinum. His repeated attempts to reach this celestial treasure by drilling were inconclusive but the area is still held as a mining claim. In 1919, an eight-inch churn drill hole was carried to a depth of 1376 feet below the south rim and is reported to have brought up numerous fragments of nickel-iron before the bit stuck in what may be the main mass of meteoritic material.

The Stone That Attracts Iron

Although he does not give it a name, Theophrastus, in his work *On Stones,* the earliest systematic treatise on minerals that has survived, describes "the stone that attracts iron" and relates it to amber as also having the power of attraction. This is undoubtedly the first statement relating the forces that we call electrostatic and magnetic. Plato mentions the stone that attracts iron and calls it "the Heraclion stone"; hence the variety of magnetite we call lodestone was known sometime before 400 B.C. Nicander (ca. 184–135 B.C.) relates the tale of the shepherd Magnes, said to be Indian or Macedonian, who first observed that the iron ferrule of his staff adhered to a certain black rock. Pliny refers to the stone variously as "Magnes siderites" and the "Heraclion stone" and accepts Nicander's story of Magnes. Later authorities assign the origin of the name to the district of Magnesia, bordering on Macedonia. In the early nineteenth century the mineral was known in England and the United States simply as magnetic oxide of iron, and the variety lodestone as natural magnet, or magnet. It became known as magnetite in 1845.

All of these names reflect magnetite's outstanding property: magnetism. Like iron, all magnetite is attracted by a magnet, but some rather rare specimens are themselves natural magnets and attract iron or smaller pieces of magnetite. This latter variety, called lodestone, was what excited the wonder and speculation of the ancients. It is doubtful that they recognized ordinary magnetite at all, although Thomas Nichols of Cambridge in his *Lapidary or History of Precious Stones,* in 1652, says, "The fourth kind is a feminine load-stone, it is black and of no use." This slander on femininity may constitute recognition of magnetite that does not possess the power of attraction but is itself attracted. Nichols also describes an earthy, bluish material with magnetic properties, probably what we recognize today as the mineral maghemite, a substance having the gross chemical composition of hematite, iron sesquioxide, while retaining the internal crystal structure of magnetite and with it, magnetite's magnetic powers.

The cause of magnetism is both compositional and structural. In order to be a magnet, a material must contain one or more of a group of elements that includes iron, nickel and cobalt. Further, the atoms of the magnetic element must be oriented in a parallel sense so that each tiny atomic magnet adds its

*The regularity of the triangular pattern brought out by etching an iron meteorite found at Atonah, Utah, shows that, except for the right end, the meteorite is a single crystal. The dark area is the iron sulfide, troilite.
(Emil Javorsky)*

effect to that of its neighbors. If the factor of parallel orientation is not satisfied, a material containing these elements may be attracted by a magnet, or placing it in a strong magnetic field may bring about the necessary orientation. This effect can easily be demonstrated with a piece of iron metal, but also obtains in materials, like magnetite, that are more difficult to orient or magnetize.

Among the countless uses of magnetic materials in our complex world, there are field magnets for radio and phonograph loudspeakers; magnetic pickups for record players; cores for radio-frequency coils used as antennas in small radio sets; magnetic deflectors and ion traps in television sets. These modern magnetic materials are essentially synthetic magnetite, and go under the name of ferrites in industry. Other metals such as magnesium, manganese, nickel and cobalt can be substituted for iron in a magnetite-like structure, thus obtaining a variety of properties for different applications.

Magnets have always been a source of wonder, especially before they were understood. Thus a medieval writer such as Thomas Nichols solemnly declares that in India magnetite was so plentiful that ships had to be built with wooden pegs instead of iron for fear the nails would be pulled from the planks! Most fanciful of all the applications of magnetite is that described by Dioscorides in his *Materia Medica* (ca. A.D. 50); namely, if a piece is secretly placed in the bed of a chaste woman, she will embrace her husband; if she is not above reproach, it will cause her to fall out of bed!

Hematite

Both Theophrastus and Pliny describe a stone, *haematitis*, as blood-red in color (hence its name) and solid in texture, and make plain that they consider it a gem suitable for cutting into seals, rings or amulets. But hematite, when it is of sufficiently "solid texture" to permit cutting as an intaglio gem—as it is used today—is not blood-red but shining metallic black. It would seem that the "haematitis" of the ancients is therefore red jasper rather than our hematite. But Pliny obviously also includes oxide of iron under the same name. It is not

Polished and etched iron meteorite found in 1959 near Bagdad, Mohave County, Arizona. (Floyd R. Getsinger)

the color of the crystalline substance that gives the mineral its name, but the color of the powder, or the streak, as it is called today. This is brownish-red, sometimes called Indian red, a familiar pigment used in paint for barns, railway cars and all applications where an inexpensive but exceptionally durable color is desired. A part of the palette of every primitive artist, hematite has displayed unequaled durability and constancy of color. Mixed with animal fat, it yielded the glowing reddish tones of the Paleolithic animal paintings in the caves of the Pyrenees. Applied to the walls of Egyptian tombs, its color is as fresh today as it was thousands of years ago. But hematite was applied not only to cold tomb walls but to living flesh as face and body decoration, and it is still the coloring constituent in the rouge on many a female cheek. Another application of finely ground hematite, or rouge, is as a fine polishing agent for metals, hard stones and optical lenses, an application in which it still has no rival.

Under favorable circumstances, hematite crystals may form in cavities, as on the Island of Elba and in the crystal caves of the St. Gotthard massif in Switzerland; it may appear along with quartz, barite and fluorite in the mines of Cumberland, England; and in the province of Minas Gerais, Brazil. These are platy hexagonal, glittering black crystals, and may have somewhat curving lustrous faces. Also found in the Cumberland district is a variety of hematite called by Nichols "Haematites pulcherrimus, or the fair Hematite, which resembleth in form a discovered brain, which form, saith Rulandus (in his *Lapidary*), I cannot sufficiently admire." In this variety, sometimes known as "kidney ore," slender fibers form radiating groups that terminate in rounded surfaces and create the likeness to a kidney or a brain remarked on by various authors. This type of hematite was also noted by Pliny, who called it botryitis, from the Greek word for a bunch of grapes. Modern mineralogists employ the adjectives botryoidal, that is, resembling a bunch of grapes, or reniform, resembling a kidney, to describe such surfaces. The silky texture of the surface, as well as the subtle shadings of the curving hummocks, make this variety especially prized by collectors.

Hematite is often earthy and ocherous when deposited under sedimentary conditions, but when crystallized by the high temperatures and pressures prevailing deep in the earth it may form glittering crystalline plates. These thin parallel plates make the rock in which they occur schistose. Because the plates are mirror-like, this variety is called specularite or specular hematite (from the Latin *speculum*, meaning a mirror). In some very coarsely crystallized specimens individual plates may be inches in diameter.

Sometimes octahedral crystals resembling magnetite in appearance prove to be wholly nonmagnetic, and when rubbed on the streak plate, yield the characteristic red powder of hematite. These crystals, called martite, are pseudomorphs of hematite after magnetite, that is, they were crystals of magnetite which by chemical alteration became hematite without change of external form. Simple oxidation could bring this about. Martite is an important ore of iron in a few localities, such as Star Lake in the Adirondacks of New York and Magnetogorsk in the Ural Mountains.

Goethite and Limonite

In one of his witty poems Arthur Guiterman writes:

> The tusks that clashed in mighty brawls
> Of mastodons are billiard balls.
> The sword of Charlemagne the Just
> Is ferric oxide known as rust.

Rust is not, alas for the meter of the verse, just ferric oxide; it is ferric oxide that contains water of crystallization. A crystallized mineral of definite properties and composition, it is named goethite for the German poet and dramatist. The name limonite, formerly used indiscriminately for most of what is now called goethite, is now properly restricted to noncrystalline, earthy, impure mineral mixtures. They often contain goethite as a principal constituent but also include other, less well-defined hydrous oxides of iron, silica, manganese and other impurities. Since the names goethite and limonite refer to purity and crystallinity, one can always identify a well-crystallized specimen as goethite, but when a specimen is earthy or amorphous-appearing, one cannot identify it by observation alone. Goethite and limonite, like hematite, are natural pigments, imparting to the rocks and minerals in which they occur yellow and brown colors.

Minerals, like Charlemagne's sword, may rust and be converted in whole or in part into goethite or limonite when exposed to the weather. Some of the iron is carried away in solution, later to be deposited in sedimentary beds, but some of it remains behind at or near the site of the original mineral as brown stains or cellular masses of goethite or limonite. The surface expression of many ore bodies is a rusty crust of goethite or limonite, called the gossan, formed as a result of the weathering of iron minerals, particularly iron sulfide, pyrite. In many places, these solutions may attack single crystals of pyrite, completely replacing them with the oxide but preserving their characteristic crystal form. These are called pseudomorphs of limonite after pyrite and are nature's most common pseudomorph.

Goethite, like hematite, forms silky fibrous aggregates with reniform surfaces, often showing a beautiful concentric color banding in shades of yellow and brown. Such specimens have been found in large quantity at Negaunee and Marquette, Michigan, and in Westphalia and the Rhine provinces of Germany, particularly at Siegen. Both velvety coatings and superb crystals have come from Příbram, Czechoslovakia.

Small, but sharp and well-formed goethite crystals are sometimes found perched on the calcite or quartz crystals lining cavities formed by solution and deposition by circulating groundwater in sandstones and limestones.

Siderite

Siderite ($FeCO_3$) is a first cousin to calcite ($CaCO_3$) and has the same internal arrangement of atoms, differing from calcite only in that iron atoms take the place of calcium. Siderite also has, like calcite, a perfect cleavage in three directions not at right angles to each other and its crystals, although rare, resemble those of calcite. It is a carbonate, and if it is attacked by acid of the proper concentration and temperature, it will effervesce and give off bubbles of carbon dioxide gas. However, because the electrical binding forces in siderite are somewhat stronger than those in the softer calcite, hot, rather concentrated acid is required to make siderite effervesce, whereas cold, very dilute acid suffices to make calcite react vigorously. The presence of iron rather than calcium has the further effect of giving siderite a yellow to brown color, whereas calcite, when pure, is white or colorless. Also, because iron atoms are heavier than calcium atoms, siderite has a specific gravity of about 3.5, whereas calcite is only 2.71. These two minerals illustrate very simply and clearly how chemical composition determines the properties of minerals, since the structural arrangement is the same for both.

Iron Sulfides: Pyrite

Pyrite is more commonly found in crystals than any other sulfide. The characteristic habit of a grooved or striated cube is alone sufficient to identify it. The other two crystal forms found on pyrite crystals, the pyritohedron, which as its name implies, is often seen on pyrite and seldom on any other mineral, is a twelve-faced form, each face of which is, in the ideal development form, almost a regular pentagon. The octahedron is an eight-faced form, all faces of which in ideal development are equilateral triangles. Rarely does pyrite form crystals

(Above) Two hundred miles east of Antofagasta on the border of Chile and Argentina the Laco iron ore deposit rises to 15,300 feet. Because of its inaccessibility this mountain of magnetite remains essentially undeveloped. (U.S. Geological Survey)

(Below) Slender crystals of quartz have grown on a background of yellow pyrite at Zacatecas, Mexico. (Studio Hartmann)

on which the octahedron is the sole form, and it is most commonly seen as small shiny faces cutting the corners from the cube or pyritohedron. Even when massive, pyrite is easily recognized by its brass-yellow color, the absence of cleavage, and its hardness—greater than that of the knife, which is unusually high for a metallic sulfide. There is no excuse for confusing pyrite with gold, which is soft and sectile, of a much deeper yellow color, and can be cut like lead by the knife.

Marcasite

With marcasite, the usual confusion respecting names rises to plague the historian, because the name marcasite was used in Europe until the late eighteenth century for what was apparently pyrite. The name pyrite derives from the Greek word for fire, in allusion to the sparks that issue from pyrite when struck with steel. The name for a time was used both for marcasite and what is now termed pyrite. Marcasite and pyrite have an identical chemical composition, iron disulfide, but have a different internal structure and hence crystal habit. Marcasite commonly forms spear-shaped crystals, or so-called cockscomb twins, very different in appearance from the cubic or pyritohedral crystals of pyrite. It is a low-temperature mineral, characteristically formed in sedimentary rocks from acid solutions. When a single chemical compound can occur in this way in two different crystalline modifications, we would expect that, as in many dimorphous pairs of minerals, only one would be stable in any given environment. However, we sometimes find pyrite and marcasite in intimate intergrowths or in oriented overgrowths one upon the other, as at Bredelar, Germany. In the iron sulfide concretions that occur abundantly in the Devonian shales of central Ohio, pyrite and marcasite occur side by side in the sulfide concretions, some of which weigh many pounds and are solidly made up of sulfides. In general, however, pyrite tends to be formed at higher temperatures and from more alkaline solutions than marcasite. Spear-shaped crystals of marcasite of fine quality have come from the plastic clays of the brown-coal formations at Carlsbad and Teplitz, Czechoslovakia. Marcasite is also a constituent of the mineral assemblage of the Tri-State ores of the region near where Kansas, Oklahoma and Missouri meet. Here crystals of galena, sphalerite, chalcopyrite calcite, dolomite and marcasite line caverns and passages in the silicified limestone that is the host rock of the deposits. Not only are these deposits important ores of zinc and lead, but they have furnished specimens of beauty and distinction to every museum of the world.

Pyrrhotite

Named for the Greek, pyrrhos, meaning reddish, pyrrhotite has a distinctive color resembling slightly weathered statuary bronze but with a creamy or pinkish quality. Pyrrhotite is unusual in that it is the only mineral other than magnetite that may show a strong response to a magnet. The mineral is commonly massive, but distinct crystals of pyrrhotite in the form of hexagonal plates are occasionally found. The chemical composition is unusual too, in that it approaches simple iron sulfide, FeS, but always displays in its analyses a deficiency of iron, so that its formula was written in the past in a great

(Above) Hematite, as shown by its characteristic color, is the ore mineral at this iron mine at Rio Marina on the Isle of Elba. (Werner Luthy: Bavaria-Verlag)

(Below) Calcite stalactite encrusted with crystals of marcasite from Charcas, Mexico. (Reo N. Pickens, Jr.)

variety of ways: as Fe_7S_8, Fe_8S_9, and Fe_9S_{10}. It has been established that all of these formulas are equally correct, or incorrect, as one may choose to view it, because the amount of iron deficiency is continuously variable and depends on temperature of deposition. The higher the temperature the less the amount of iron and the more magnetic the pyrrhotite!

The Iron Ores

The ores of iron are oxidized compounds in which iron is bound to oxygen by strong electrical forces that must be broken to obtain the metal. This operation, called reduction, is carried out in a blast furnace and requires a great deal of energy. Thus in addition to an ore, it is necessary to have a fuel of charcoal or coke, and a flux such as limestone. The furnace is charged with these ingredients in the proper proportions and the ignited fuel under forced draft heats the mixture to a high temperature. Under these conditions, iron is liberated from the combined oxygen and becomes molten while the limestone combines with impurities, usually silica, to form a fluid slag. The location of the world's iron smelting centers have been determined by many things including political boundaries, transportation facilities, and centers of consumption, but chiefly by proximity to ore, fuel and flux. Ideally, all three should be in the same area.

The disappearance of the world's forests may be largely attributed to the vast quantities of wood burned for charcoal to be used in the manufacture of iron. It was not until the eighteenth century in England that an iron industry using coke as a fuel emerged on a large scale.

Magnetite Ores

Magnetite is a widespread mineral most commonly found in tiny crystals scattered through igneous and metamorphic rocks. Even though the total amount of such magnetite is tremendous, the concentration at any one place is insufficient to constitute an ore. Magnetite is sometimes one of the first minerals to crystallize from a cooling magma, and its high density, much greater than the surrounding melt, permits it to settle slowly and accumulate at the bottom of the magma chamber. From some magmas ilmenite and magnetite crystallize at the same time, and the accumulation is a mixture of the two minerals. This process of magmatic segregation has taken place in many parts of the world to form workable iron ore deposits. But the largest such magnetite accumulations are in northern Sweden, with the most noteworthy at Kiruna.

In this land of midnight sun, well above the arctic circle, stand the twin ore mountains of Kiruna. The larger, Kiirunavaara, is separated by a lake from Luassavaara. Since mining began there in 1903, over 300,000,000 tons of ore have been mined but an estimated reserve of over 1,300,000,000 tons remain. The ore is magnetite, with the only impurity small amounts of irregularly disseminated apatite. These sill-like tabular bodies of essentially pure magnetite presumably were injected as a liquid along the contact between two igneous bodies; it then solidified in place. Mining originally took place on the surface, but it now goes on underground so that operations can continue the year round despite the long cold winters. Ore from Kiruna and other Swedish iron deposits is more than adequate to supply the blast furnaces of Sweden, and much of it is exported. To reach world markets the ore is shipped by rail to the seaport at Narvik, Norway, and thence by ship.

The industrial need for iron was demonstrated during World War II. For many years before that much Swedish ore had been shipped to Germany, and a continued supply was essential for the German war effort. More to deprive Germany of the ore than to obtain ore for herself, England determined to stop the flow of this vital material. The result was a great naval battle in and around the port of Narvik with heavy losses on both sides. Germany remained in control of the port, but air strikes and the operations of Norwegian partisans prevented the Germans from maintaining shipments.

Magnetite was the chief ore mineral of the infant iron industry in the United States. Numerous though small deposits were worked in Connecticut, New Jersey and Pennsylvania until the middle of the nineteenth century when rich hematite ores were discovered in the Lake Superior region. Now, one hundred years later, new magnetite deposits are being developed deep beneath the rolling farmland of Pennsylvania. These and other hidden ore bodies in many places in the world have been located by means of a magnetometer, an instrument that can detect minute differences in the magnetic field of the rocks due largely to magnetite. A high "magnetic anomaly" indicates the presence of a buried magnetite deposit.

Ilmenite Ores

Ilmenite, an oxide containing both iron and titanium, takes its name from the Ilmen Mountains in Russia. It is a hard, black mineral frequently confused with magnetite since it is found in similar deposits formed under the same conditions; it is, however, nonmagnetic. In fact, the two minerals frequently crystallize together in such an intimate aggregate that they must be powdered to separate one from the other. In the early days of iron smelting, such mixtures (titaniferous iron ores), assumed to be magnetite, puzzled operators, for the ore did not react properly in the blast furnace and proved unusable.

A large deposit of magnetite-ilmenite associated with gabbroic rocks in the Adirondacks at Sanford Lake, New York, has been known since 1836. After futile early attempts to recover iron from it, it was abandoned until modern ore dressing methods pointed the way to a separation of the two minerals by magnetic separators. It is now mined on a large scale chiefly for the titanium, since the magnetite fraction still contains some ilmenite, making it poor iron ore. From the ilmenite comes titanium dioxide, widely used as a white paint pigment, but even more important, as the principal ore of titanium metal.

Titanium is a modern metal and exploration of outer space will result in part from its unusual properties. Although it is sixty per cent heavier than aluminum, it is so much stronger that a structure requiring a certain strength is lighter when made from titanium than from aluminum. Further, it has a high melting point 1800° C (3272° F) and is more corrosion-resistant than stainless steel. Because of these properties its list of uses, now long, will surely increase with greater production and lower price. The bulk of the metal is now used in aircraft frames, jet engines and missile and space components. When man eventually lands on the moon he will arrive there in a vehicle with many vital parts made of this exciting new metal.

Hematite

If a single mineral were selected as the most important to our industrial civilization, it would be hematite. Although it contains slightly less iron than magnetite its widespread occurrences in large mineable deposits makes it the more important ore. As we have seen, it is one of nature's major pigments, imparting its color to many rocks and to other minerals. Thus the red rocks of the Painted Desert and of the Grand Canyon, as well as the more delicate shades of the quartz varieties, carnelian and agate, result from the presence of small amounts of hematite. But only when relatively pure hematite has accumulated in deposits measured in millions of tons does it become an ore of iron.

Although hematite can form in many ways, the largest deposits are of sedimentary origin. Most of the iron liberated by the weathering of rock-forming minerals is dissolved in the surface water and eventually reaches the sea. Indications are that much of this iron has been deposited in shallow seas, sometimes covering hundreds of square miles, in the form of hematite or goethite. The reasons for the formation of hematite in some places and goethite in others is puzzling and inadequately understood. The chief impurity in iron ores is silica, indicating that finely divided quartz was deposited contemporaneously with the iron mineral. Many of the large sedimentary iron ore deposits are associated with the oldest rocks, the pre-Cambrian, indicating that this process has been going on throughout geologic time. Some of the ancient deposits have become metamorphosed as in Brazil and Michigan; the resulting recrystallization has converted the red hematite into an aggregate of black micaceous scales called specularite. In more recent rocks such as the Clinton iron ore beds of Alabama the marine character is indicated by the fossils they contain.

Major hematite ore bodies too numerous to mention are found on every continent and their exploitation is constantly being accelerated. With the refinement in geophysical prospecting in recent years, many new deposits have been located in remote parts of the world.

Since 1895 the United States has been the world's largest consumer of iron ore and until 1960 the largest producer as well. Until the middle of the nineteenth century the chief American source of iron ore was small magnetite deposits in the eastern United States. At that time hematite was discovered in "inexhaustible amount" on the iron ranges around the southern and northwest shores of Lake Superior. Of the ranges there the Mesabi is the giant, and since 1892 has yielded over 2,500,000,000 tons of high-grade ore, over twice the total production of all the other ranges.

In 1854, these Western iron ores were a long way from centers of consumption and of coal for the fires of the blast furnace. But they were transported by barge or freighters from Duluth through Lakes Superior and Huron to the southern shore of Lake Erie. A relatively short haul by rail brought them to Pittsburgh where there was an abundant supply of excellent coking coal and limestone for flux. Although other steel-producing cities appeared south of Lake Erie and later on Lake Michigan, this flow of iron ore remained basically the same for one hundred years.

In the 1940's the producers began to see the end of the "inexhaustible" high-grade ores of the iron ranges. Since that time new districts have been developed, mostly outside of the United States. The greatest is in Canada in the Ungava trough on the Labrador-Quebec border. From here hematite ore is brought by rail to the St. Lawrence River and floated through the St. Lawrence Seaway to ports on the Great Lakes. Deposits in western Canada, yet undeveloped, are reported to be of tremendous potential reserve. In South America an expanding production from huge deposits in Venezuela, Brazil, and Chile is supplying high-grade hematite ore to the Western world, much of it to the United States. Expansion of the iron ore industry is world-wide; production figures from Australia and twenty countries in Africa and Asia show increases each year.

In addition to importing high-grade iron ore to supplement its dwindling reserves, United States companies have turned their attention to the primitive iron formation called *taconite* in which the high-grade ores of the Lake Superior iron ranges lie forming a small although rich part. Taconite is low-grade compared with high-grade "shipping ore" and contains only twenty-five to thirty per cent iron. However, it covers hundreds of square miles and the reserves are colossal. Today taconite is being mined on a large scale and millions of tons of nearly pure magnetite and hematite are being separated from it. Although the process is complicated and costly, this "manufactured" ore contains as much as sixty-five per cent iron and can be considered a reserve for the indefinite future.

Goethite and Limonite Ores

In the normal process of weathering, much of the iron resulting from the chemical breakdown of the rock-forming minerals is carried away in solution. Most of this iron in solution reaches the sea, but some of it is carried into marshes and bogs. Here, in mainly stagnant water, the iron may oxidize to an iridescent surface film with the aid of iron-using bacteria and sink to the bottom. In this way deposits of goethite-limonite called "bog iron ore" can be built up. It was from such bog ore, recovered from the marshes near Saugus, Massachusetts, in the 1640's that iron was first produced in the New World.

Under certain tropical and subtropical conditions the chemical reactions that take place during rock weathering differ from those in more temperate climates. Instead of being transported in solution, iron forms hydrated insoluble oxides and remains on the surface while other major elements of the rock are removed. Aluminum reacts in a manner similar to iron and vast areas of the earth's surface in tropical regions are covered by oxides of these two elements. Such deposits are laterites. If the original rock is rich in aluminum, the resulting laterite is bauxite, the ore of aluminum. If, on the other hand, the weathered rock is iron-rich and aluminum-poor, a surface concentration of goethite-limonite results. At Mayari and Moa in Cuba extensive iron ores of this type have been formed by the weathering of the iron-rich rock serpentine.

The greatest accumulations of goethite-limonite—less extensive than those of hematite—were produced in the same manner, that is, by deposition from the sea. This partly chemical and partly biochemical process has in places produced vast tonnages of "brown iron ores." The most noteworthy occurrence of such a deposit of goethite is the minette iron ore of Alsace-Lorraine and Luxembourg. These deposits, with estimated reserves of five billion tons, lie

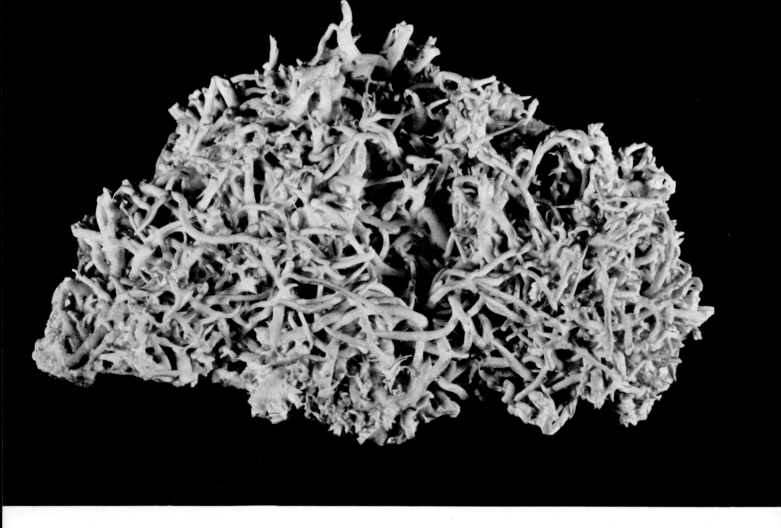

near the border of France and Germany and have contributed to the industrial prosperity of the two countries; at the same time they have for a century been a source of conflict between them.

Siderite

Although iron carbonate, siderite, is a common mineral with a world-wide distribution, it is only a minor ore of iron. It is most commonly found in small amounts in veins associated with ores of lead and silver. In some large veins, as at Westphalia, Germany, it may be the dominant mineral, and is used as an iron ore. The Austrian deposit at Erzberg in Styria is the only concentration of siderite of sufficient size and purity to be considered a major source of iron. Here a folded limestone has been replaced by massive siderite through the agency of iron-bearing waters to form a deposit of many tens of millions of tons.

Pyrite Ores

Pyrite forms in nodular masses on the sea floor at near freezing temperatures. At a temperature of hundreds of degrees it crystallizes from molten magma

When aragonite in coral-like aggregates is found on the walls of iron mines it is called "Flos Ferri," meaning iron flower. *This specimen is from Styria, Austria, which is famous for this variety of aragonite. (Cornelius S. Hurlbut, Jr.)*

as an accessory mineral in igneous rock. And at all temperatures between these extremes, it forms as the most abundant metallic mineral in ore veins. Because of the wide range of the conditions of its formation and its universal occurrence, unmatched by any other mineral save quartz, pyrite (iron sulfide) has thus been called a "persistent" mineral.

Most properly, pyrite should not be considered with the ores of iron, for notwithstanding its wide occurrence, the many large deposits, and its high percentage of iron (46.6), it is rarely used as an iron ore. Nevertheless, it is mined not as an ore of iron but an ore of sulfur. When pyrite is roasted, that is, heated in the absence of air, its sulfur, amounting to 53.4 per cent in the pure mineral, is driven off and recovered by solution in water as H_2SO_4. In addition to pyrite, but to a much lesser extent, sulfur is obtained from the other iron sulfides, marcasite and pyrrhotite. These three minerals go under the trade name of *pyrites* and jointly yield nearly forty per cent of an annual world sulfur production of 21,000,000 tons. Native sulfur is the major source of the element.

The chief pyrite-producing countries in the Free World are Japan, Spain, Italy and Norway. Probably the largest reserves are in the Rio Tinto district of southern Spain and adjacent parts of Portugal. Here, in a belt eighty by twenty miles are irregularly distributed, gigantic lenses of nearly pure pyrite varying in length from twelve hundred feet to over a mile. These deposits, each containing tens of millions of tons, were known and mined by the Romans and even earlier by the Phoenicians. Their rather extensive workings were not undertaken for sulfur but for the small amounts of copper and gold associated with the pyrite.

14 Minerals of the Precious Stones

Through the ages, men have used mineral products as decorations in their homes, as adornments for themselves or their gods, as symbols of their status and as portable wealth. Of the many minerals so employed, only five, the diamond, the ruby, the sapphire, the emerald and the opal, have achieved such a reputation that they merit the title of "precious stones." The pearl might well be added to this distinguished group but it is a product of organic nature and not strictly a mineral.

All of these stones, with the exception of the opal, have been produced not only by nature but as synthetic products by man. The haunted history of the opal, that mysterious gem whose siren beauty results from a lack of ordered internal structure, is discussed elsewhere. The remaining four precious stones are today synthesized in quantity as profitable commercial commodities. The synthetic stones are indistinguishable from the natural stones in most of their physical and chemical properties. They are not "fakes" or "imitations" but the very same chemical element or compound that constitutes the natural minerals. The story of the development of the methods of synthesis is most significant because it reflects the change in man's relations with his natural environment: once he was a helpless part of it; today, he in large part dominates it.

The Diamond

Of nearly two thousand mineral species the diamond is better known than any other. Not only is it the most coveted gem stone today, but it has been so throughout history although it was rare among the ancients. Pliny in A.D. 100 writes of it as "the most valuable of gem stones, but known only to kings."

The high value that man has always placed on the diamond results from its

The "Big Hole" of the Kimberley diamond mine, with the city of Kimberley, South Africa, in the background. Between 1872 and 1914, 25,000,000 tons of rock yielded 14,504,566 carats of diamond. (Tad Nichols)

remarkable physical properties. It is above all hard, harder and more resistant to abrasion than any other mineral; nothing will scratch it except another diamond. Moreover, the diamond is unsoluble in all acids and alkalis. Because of these resistant properties the name *adamas* from the Greek meaning "the invincible" was early given to the mineral and the present name *diamond* is derived from this. Although diamond has a great hardness, there are four directions along which it can be cleaved and thus easily broken if struck a sharp blow. This property of brittleness may have resulted in the destruction of many fine stones. It is reported that the early explorers in South America understood the diamond to be invincible in all regards—tough as well as hard. Their test was to place the suspected stone on an anvil and strike it with a hammer. If it resisted this harsh treatment, it was a diamond; if it was crushed to a powder it didn't matter since it was thus revealed to be a worthless mineral.

The chemical composition of diamond is extremely simple; like graphite it is composed only of the element carbon. But there the similarity between these crystalline forms of carbon ends, and no two minerals have a greater diversity of properties. Diamond is hard, lustrous and transparent; graphite is soft, dull and opaque. Diamond has a specific gravity of 3.5, high for a nonmetallic mineral, and thus tends to collect in placers and alluvial deposits. Graphite's specific gravity is 2.2, extremely low for a metallic mineral. These strikingly different properties of the same element result from the manner in which the carbon atoms are packed together. In diamond they are close together and held by strong electrical bonds, whereas in graphite they are far apart and have weak bonds between them.

The luster of the diamond is more brilliant than of most other minerals and even an isolated stone in a stream catches the eye of the prospector. It also has an unusually high dispersive power that gives rise to the brilliant flashes of spectral colors one sees when white light is refracted through a cut stone.

Most people are familiar with diamonds only as "white" or colorless, unaware that these stones occur in a great range of colors. The greatest variety of shades is found in yellow and brown; rarer are green, red, blue and violet. Carbonado is a black granular variety; it is tough as well as hard and not easily broken.

Only about twenty per cent of the diamonds mined are white or blue-white and of sufficiently high quality to be cut as gems. The others, off color or flawed, are called *bort* and are used for industrial purposes. A stone of gem quality but with a slight tint does not command as high a price as an otherwise similar but colorless stone; however, transparent stones with a deep color, such as the famous yellow Tiffany diamond and blue Hope diamond, are of far greater value. Stones of high quality that are a deep red, blue, yellow or green called "fancy stones" are greatly prized and, like art objects, have no fixed value.

Diamonds differ markedly in color, average size and quality from place to place. At one locality, most of them may be of gem material and colorless; in another nearly all may be tinted yellow or brown, have flaws and be useful only in industry. Nevertheless, the two sources will have one feature in common—a high percentage of the diamonds from both will occur in individual crystals. The rock in which the diamonds grew permitted equal development on all sides of the stone, giving rise to well-formed crystals. Because of their

(Above) The final mining operation at the Consolidated Diamond Mine, Southwest Africa. After removing thirty to forty feet of sand, the workers probe cracks in the bedrock and sweep the surface in their search. (Siegfried Muessig)

(Below) Mining rubies from the stream gravels of Burma still follows the ancient hand methods of digging, screening and sorting. (Reginald Miller)

extreme hardness, the original crystal forms are preserved even though a diamond has been transported as a stream pebble for hundreds of miles during tens of thousands of years. Some crystals are dodecahedral, others cubic and still others nearly spherical and covered with many faces, but the most common are octahedral, that is, with eight faces as two four-sided pyramids base to base.

Diamonds were first found in India, this remained virtually their only source until they were discovered in Brazil in 1725. In ancient times diamonds came mostly from stream gravels and alluvial deposits over a wide area of southern and central India. Gathered one by one from many localities, the stones were brought for marketing to the town of Golconda. Although no mining was done at Golconda itself, the name has become synonymous with a rich mine or source of great wealth. It is estimated that during the several thousand years that India was the sole source of the world's diamonds, twelve million carats were produced there. Many of the most famous stones such as the Kah-i-nor, the Great Mogul and the Orloff are of Indian origin, and each has a long history of intrigue, theft and murder.

In Brazil, diamonds were discovered in the state of Minas Gerais near the present town of Diamantina. While working the stream gravels for gold, slave miners were attracted by bright pebbles in the gold concentrates; some of these found their way to Portugal and were recognized as diamonds. The search that followed located diamonds in the states of Bahía, Goyaz and Mato Grosso. At first Brazilian diamonds were not accepted by Western buyers, and for a while they had to be sent to Golconda before they could be sold in European markets. But by that time Indian production was waning fast and Brazil soon became the world's principal source of these precious stones. After nearly two and a half centuries, Brazil is still a producer of gem diamonds of high quality.

The next major diamond discovery was in 1866 in South Africa. In that year a bright pebble picked up by the son of a Boer farmer from the bank of the Orange River eventually reached Dr. Atherstone, a mineralogist in Grahamstown, who recognized it as a diamond. It weighed 21½ carats, and sold for £500. A year later other stones were discovered on the banks of the Vaal River. But it was not until 1869, when a native shepherd boy picked up a brilliant of 83½ carats, that attention was drawn to diamonds in South Africa. Adventurers poured into the area from the world over and a rush began that was comparable to the California gold rush of 1849. Soon ten thousand men were digging in the gravel for over one hundred miles along the Orange and Vaal Rivers. Some of the more adventurous prospectors tried sifting the soil on high ground and were rewarded by finding diamonds there also. At the surface these diamonds were found in loose, unconsolidated soil not unlike those in the river gravels. But as the workings deepened, the diamonds of the "dry diggings" were found embedded in a soft yellow rock called "yellow ground." At a depth of fifty to sixty feet this friable "yellow ground" gave way to a harder and more dense "blue ground," but the diamonds continued. For the first time diamonds were being recovered from the rock in which they formed. It was a volcanic rock occupying a great tube called a "diamond pipe" that had pushed its way upward from deep in the earth and formed a circular area at the surface. The rock in which the diamonds are embedded is known today as *kimberlite,* a

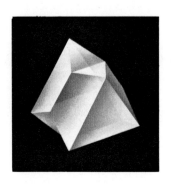

Diamonds are commonly found in well-formed crystals even after they have been ground in the sands of a stream for thousands of years. The most common crystal form is the eight-sided octahedron (above), but almost as abundant are twins composed of two half-octahedrons joined together in reverse positions (below).

name taken from the most famous of the diamond pipes, the Kimberley.

Although several pipes were soon discovered, the Kimberley was the richest and therefore its exploitation progressed most rapidly. Where small claims 30 × 30 feet had proved feasible for working river gravels that rarely went deeper than fifteen or twenty feet, this was most impractical for mining rock that extended indefinitely downward. As the digging went deeper the roads between the claims collapsed, slides from one claim to another resulted, and by 1886 the caving in of the sides of the ever deepening crater brought an end to individual enterprise at the Kimberley pipe. In 1889 Cecil Rhodes took over many small claims to form the DeBeers Consolidated Mines, Limited, and diamond mining became big business. Shafts were sunk beyond the margin of the pipe from which tunnels were driven inward to the diamond-bearing blue ground. When operations ceased at the Kimberley pipe in 1914, mining had reached a depth of 3500 feet, 25,000,000 tons of rock had been excavated, and 14,504,566 carats, or three tons, of diamond had been recovered. Today, other diamond pipes nearby are still major producers of diamonds and the city of Kimberley remains a center of the diamond industry.

The proverbial needle in the haystack is not more difficult to find than a diamond in blue ground. At the Kimberley pipe there was eight million times as much barren rock as diamond; elsewhere the ratio is as high as 1:30,000,000. It is thus possible for a miner to work a lifetime in a mine without finding a diamond. A major problem in the early days of underground mining was to prevent miners from stealing the stones in the final hand-sorting process. Quite by accident, a method was discovered that almost eliminated the human element. It was noted that of the minerals in the wet concentrate only diamonds stuck to grease; the worthless minerals were washed away. Therefore, today the

Sifting Diamond Gravels, *an illustration appearing in* Scribner's Monthly, 1873. *The early digger in South African diamond mining usually had several natives working for him, but the digger himself sifted the concentrate. (Historical Pictures Service-Chicago)*

final process is to wash the concentrate across tables coated with grease. This not only reduces the opportunity for theft but catches the small diamond that the most honest workman might not detect.

The normal work contract for a native is for one year. Then he is sent home even though he may later return. It was not an uncommon practice for the worker just before his discharge to swallow diamonds he had discovered. These he later recovered and sold. Now each departing worker is x-rayed to detect any stones he may have swallowed.

By the end of the nineteenth century prospectors could recognize diamond pipes from certain surface features. One such man was Percival White Tracey, who had worked a claim at Kimberley. Near Johannesburg he found diamonds in a stream, and following the stream toward its source he came, twenty-four miles east of Pretoria, on what appeared to be a diamond pipe. The owner of the land, one Joachim Prinsloo, having been persuaded to sell two earlier farms to mining men, would permit no trespassing on his land or sampling of the soil. But Tracey was so sure that diamonds lay beneath its surface that, without sampling, he entered into negotiations for the land. The price, £55,000, was very high, but with the aid of Thomas Cullinan, a building contractor, Tracey raised the money, bought the land and formed the Premier Diamond Mining Company Limited.

Mining began in April, 1903, and continued until 1931, producing in that time thirty million carats. In 1932 production ceased and the "Great Hole" filled with water. But in 1944 the Premier mine was reopened and, with modern mining methods, became again one of the largest diamond mines in the world, with an annual production of over two million carats.

For many years the Premier was not only the largest diamond mine but had the distinction of yielding the largest diamond ever found. In 1905, an African

This illustration of washing the river gravels in South Africa for diamonds appeared in Scribner's Monthly *in 1873. Although some individual diggers still work the gravel in this primitive way, most diamond mining today is big business. (Historical Pictures Service-Chicago)*

worker pointed out a shiny object on the wall of the open cut to the surface manager, F.G.S.Wells. Mr.Wells picked from the rock a diamond weighing 3024¾ carats! This diamond, named the Cullinan after the chairman of the company, was purchased by the Transvaal Government for £150,000 and presented to King Edward VII. It was cut into nine large stones and ninety-six smaller ones. The largest, the "Star of South Africa," was mounted in the British royal sceptre and may be seen in the Tower of London.

Five years after work began at the Premier mine, diamonds were discovered in German Southwest Africa. The diamonds lay on the polished rock surface of an arid, desolate region from which winds had removed any soil cover. "Mining" consisted of a line of natives who picked up diamonds as they crawled on their hands and knees, elbow to elbow.

In 1927 diamonds were found in the sands of ancient sea beaches south of the mouth of the Orange River in Namaqualand. Then Consolidated Diamond Mines prospectors discovered diamond-bearing gravels associated with shorelines and marine terraces for more than fifty miles north of the river. The mining operation that followed was a gigantic sand-moving operation, for the diamond-bearing gravels lie beneath a layer of sand from thirty to sixty feet deep. The sand is removed and only the last three or four feet of gravel overlying the bedrock is sent to the diamond recovery plant. After the sand and gravel have been cleaned away, the bedrock is swept and every crevice in the underlying schist is probed. This unique mining procedure produces 100,000 carats a month; and since eighty per cent are of gem quality, the Consolidated Diamond Mine is by far the world's largest producer of gem diamonds.

The generally accepted theory concerning the Orange River diamonds is that they originated in diamond pipes near Kimberley, a thousand miles away, and were transported down the Orange River, discharged into the Atlantic Ocean and swept onto the beaches by littoral currents. If this is true, gravels on the sea floor should also contain diamonds. In 1962, initial operations did indeed prove that diamonds could be recovered from beneath the sea and barges and dredging equipment are today scraping diamonds from the sea floor off the coast of Southwest Africa.

All but a small fraction of the world's diamonds now come from the African continent. Alluvial diamonds were found in the Congo in 1903 and production has steadily increased until today nearly sixty per cent of the world's supply is won from the stream gravels of that area. The value of the Congo diamonds is not in proportion to the amount mined, since only five per cent are of gem quality. Other major sources are Ghana, Sierra Leone, and Angola, and to a lesser extent Mauritania and the countries formed from former French Equatorial Africa.

With one major exception all the African diamonds mined outside of South Africa are alluvial. This exception, the Williamson diamond mine at Mwadui in Tanzania, is a diamond pipe covering 347 acres. The discovery of this remarkable deposit in 1940 culminated six years of prospecting by a geologist, Dr. John T. Williamson. This diamond pipe proved as rich as it was large and in the twenty years following its discovery produced over a ton of diamonds from eighteen million tons of rock. The most famous stone yet taken from Mwadui is a pink diamond, weighing 54½ carats when found in 1947. It was presented to Queen Elizabeth as a wedding gift. As a cut stone of 23.6 carats set in a floral brooch it now forms one of the world's loveliest jewels.

Diamonds are no longer produced in North America. In the past an occasional stone turned up in the glacial drift of the Midwest, in the dunes on the southern shore of Lake Michigan, and in stream gravels being panned for gold in North Carolina and Virginia. Later, similar finds were made in the gold-bearing gravels of California. The only mining in the United States specifically for diamonds began in 1906 near Murfreesboro, Arkansas, when J.W. Huddleston found diamonds in a rock resembling the blue ground of the Kimberley pipes. This was the first discovery outside of Africa in which diamonds were found in the rock in which they grew. Over ten thousand stones, some of fine quality, were recovered but, unable to compete with the lower costs in Africa, mining ceased after twenty years. Today the area is a tourist attraction. For a small fee one can wander over the exposed kimberlite of the pipe searching for diamonds. The number of stones found is small but large enough to lure others to Murfreesboro.

There is today some production of diamonds in the Soviet Union. Although the reports are vague, the diamonds apparently come from both alluvial workings and diamond pipes in Siberia. The ratio of industrial to gem stones is also uncertain but some gems reach the free world through the Diamond Trading Company, which markets virtually all of the world's diamonds.

Diamonds for Industry

Only twenty per cent of diamonds are suitable for cutting into gems; the other eighty per cent are used for industrial purposes. Although far less glamorous, the industrial stones occupy an important place in modern technology because of their hardness.

The highest grade of industrial stones is used in the manufacture of high-precision machine tools used in lathes and drills and for trueing up abrasive wheels. Many are used in diamond drills, an indispensable tool in mining. The drill crown, studded with diamonds, is rotated on the end of a hollow rod, slowly cutting a hole through the hardest of rocks. The central core is thus left intact and is periodically broken off and brought up for inspection. In this way the miner obtains a sample of the rock deep in the earth. Before a deep shaft is sunk to an ore body, the ore is sampled by many miles of core drilling. Diamond drilling is also used to sample the ore ahead of mining before driving expensive tunnels, in exploring geologic structures and in testing the sites for dams and large buildings.

Industrial diamonds of poorer quality are crushed and sorted into various sizes for use in grinding, cutting and polishing hard materials. Small diamond saws, consisting of disks an inch or two in diameter, with diamond-impregnated rims, serve to cut gems while gigantic wheels many feet in diameter are used for sawing through granite blocks. There is a fascination in watching a spinning diamond blade slice quickly through a quartz crystal so hard that steel will not scratch it.

For centuries men have dreamed of making diamonds and during the past century there were many fraudulent claims of such a synthesis. Then, in 1955, the General Electric Company announced that it had succeeded in manufacturing diamonds that had withstood the most rigorous tests. Since that time,

(Above) Platy crystals of ruby (corundum) from Tanzania are embedded in a matrix of green zoisite. (Studio Hartmann)

(Below left) A hexagonal ruby crystal from Burma shows striations at a sixty-degree angle to each other, a common pattern in rubies. (Benjamin M. Shaub)

(Below right) An octahedral diamond from South Africa embedded in kimberlite, the rock in which it grew. (N. W. Ayer & Son)

diamonds have been made in increasing amounts, several million carats each year competing with the natural mineral. So far, the stones have been small, mostly of poor color and used only for industrial purposes. But the day may well come when gems of fine quality are produced by man at the same intense temperatures and pressures at which natural stones form deep within the earth.

Ruby and Sapphire

Although a red ruby and a blue sapphire bear no outward resemblance to each other when cut as gem stones, they are both color varieties of the same mineral, corundum. They are chemically and physically the same, differing only in the kind and amount of impurities they contain. Paradoxically, it is impurities that give these gems their value, for pure corundum is water-clear. A minute amount of chromium is present in ruby, and of titanium in sapphire.

Common corundum (aluminum oxide) is not a rare mineral; many tons of it are mined each year. Its extreme hardness, exceeded only by diamond, makes it useful as an abrasive material and contributes to its value as a gem. It permits the stones to take a high polish, one that years of ordinary wear will not dim. The hardness results from the close packing of the atoms of aluminum and oxygen in the crystal structure and the strong electrical forces that hold them together. For the same reasons the mineral has a specific gravity of 4.0, much higher than the average of nonmetallic minerals. It is this high specific gravity that causes the stones to collect in the placers where many of them are mined.

When found in crystals, corundum occurs in six-sided prisms or pyramids reflecting the internal hexagonal crystal structure. In some localities these crystals during their formation incorporate needle-like impurities of other minerals, usually rutile. As it crystallizes, corundum orients these rutile needles in three directions at angles of 60° to each other. Such crystals have the property of asterism, and in a smooth rounded *cabochon* cut, light is reflected from the rutile inclusions in the form of a six-rayed star, giving rise to the star ruby and star sapphire.

The value of a ruby depends on the color; a deep-red, known as pigeon-blood, is prized. Flawless stones of this color are usually small, rarely exceeding three carats; large rubies have been reported but the present whereabouts of most of them is unknown. Tavernier, the French traveler of the seventeenth century, described two rubies owned by the King of Bijapur, India, as weighing 50¾ and 17½ carats. Emperor Rudolph II of Germany is said to have possessed a ruby the size of a hen's egg. Other red stones such as tourmaline and spinel have been confused with rubies and it may be that the disappearance of many large rubies reported at an earlier time has resulted from the discovery that they are not rubies.

Since the fifteenth century, and possibly even earlier, the finest rubies have come from Burma, most of them from an area of about forty square miles around Mogok. The gems were formed in a marble, a metamorphosed limestone, that underlies much of the region and makes up several mountain ranges. For countless centuries rain water has been slowly dissolving the limestone, liberating the rubies to become part of the soil on the mountain sides or washing them into the valleys below. Although mining of the parent rock has produced

(Above) Emerald crystals from cavities in a gray limestone at Muzo, Colombia, source of the world's finest emeralds. (Studio Hartmann)

(Below left) The Star of India, in the American Museum of Natural History, New York, weighs 563 carats and is the world's largest star sapphire. (Courtesy of the American Museum of Natural History)

(Below right) Blue corundum, sapphire, in the matrix; and water-worn crystals of yellow and blue sapphire from Ceylon. (Studio Hartmann)

some fine stones, most rubies since earliest times have been recovered from the soil on the slopes and from the gravels in the valleys. Although modern mining methods have been introduced, some rubies are still recovered by the ancient method of digging pits and trenches.

During the metamorphism of the limestone that gave rise, geologically, to rubies, other minerals including spinel were also formed. Spinel, like diamond, commonly occurs in octahedral crystals and this enables the trained eye to distinguish them from the hexagonal crystals of ruby. However, the red transparent variety, called "ruby-spinel," if not in crystals so strongly resembles ruby that the two minerals have long been confused. A famous gem stone, the "Black Prince's Ruby" was determined to be a spinel long after it was set in the English crown. Undoubtedly many other "rubies" in fine pieces of jewelry are in fact spinel.

Rubies are also mined from alluvial deposits north of Bangkok in Thailand, and to a lesser extent at Battambang in Cambodia. Although the country is underlain by a limestone similar to that of the Mogok district in Burma, the gems are all alluvial and none has been found in the rocks in which they formed. The rubies of Thailand and Cambodia are darker and of poorer quality than the Burmese stones. Other rubies of relatively poor quality have long been recovered from the stream gravels of Ceylon, where they are associated with more abundant sapphires. Many of these are found in their original crystal form but others have become rounded pebbles.

Since ruby and sapphire are the same chemical compound and are formed under the same conditions, it is not surprising that they occur together in nature. The alluvial deposits near Battambang in Cambodia are noted chiefly for fine sapphires of cornflower-blue—the color most sought after in this gem. In the stream gravels of Ceylon sapphires are not only associated with rubies but with the other gem minerals—topaz, amethyst, garnet, zircon, spinel and tourmaline. The most productive district is in the south of the island, near Ratnapura. In spite of the name, which means "City of Rubies," sapphires are the most abundant gems here, but they are not of high quality.

In 1881 sapphires were found in Kashmir high in the Himalayas when a landslide laid bare the rock containing the stones. The stones were first recovered from the bedrock but later from concentrations in stream gravels at lower elevations. The color of these stones, known as Kashmir blue, is somewhat lighter than those of Thailand and Cambodia but it is pleasing and distinctive.

Several color varieties of corundum have been found in the United States. The earliest discovery was at Corundum Hill, in Macon County, North Carolina, where rubies predominated but sapphires were also found. In 1865 gold miners panning the gravels of the Missouri River in Montana found sapphires and other color varieties of gem corundum and many stones were taken out as a by-product of the gold mining. Sapphires were found in 1896 on the eastern slope of the Little Belt Mountains in Montana. Here, at Yogo Gulch, cutting through the underlying sedimentary rocks, is a dike of igneous rock several miles long and ten to twenty feet wide. Sapphires, most of them small but very many of fine color, were disseminated through the dike rock. At the surface the igneous rock had weathered to a claylike material that could easily be worked, but at a depth of ten to twenty feet it became harder. Thus the

(Above) A transparent octahedral diamond crystal from the Kimberley Mine, South Africa. *(Alfred Ehrhardt)*

(Above right) The Cullinan diamond, found at the Premier Diamond Mine in 1905, weighed 3024¾ carats and was the largest gem diamond ever found. It was cut into nine major stones and ninety-six lesser ones. The one shown here, Cullinan II, is the second largest stone, weighing 317 carats. *(Studio Hartmann)*

early surface operations were followed by mining from a shaft that reached a depth of two hundred feet. For economic reasons mining ceased in 1929 although untold numbers of sapphires still remain. Today one can trace the dike for long distances by following the depressions formed by the caving of the early surface workings.

In addition to ruby and sapphire, many other varieties of gem corundum are distinguished only by their color. These are formed under the same conditions as ruby and sapphire and are found in the same areas. All stones other than the red are named sapphires and according to their color; for example, yellow sapphire.

Such color varieties are also frequently given the name of another gem of similar color but with the term "oriental" as a modifier; thus "oriental topaz" is yellow and "oriental amethyst" is purple. These "oriental stones" have a higher luster and greater hardness and are thus better gems than other minerals of similar color.

Because of the high value of precious stones, man has long tried to make them artificially. His major success was early in the twentieth century with the manufacture of rubies and sapphires. Earlier, small fragments of natural ruby had been fused together to form stones, called reconstructed rubies, that could be cut into gems of several carats. But in 1902 a French professor of chemistry, A. Verneuil, announced a process by which he could manufacture both rubies and sapphires starting with a pure aluminum oxide powder. This finely divided material was passed through a hydrogen-oxygen flame, and fused into droplets at the high temperature of 3720° F (2050° C). The molten alumina dropped onto a support beneath the flame, where it crystallized, forming a pear-shaped mass called a boule. Pure alumina powder yields a colorless boule; small amounts of chromium oxide are added to produce a ruby, and iron and titanium oxides are added to make a sapphire. Other chemicals yield boules of many different colors.

The corundum produced by the Verneuil process is identical with the natural stone in all its chemical and physical properties. It is impossible for the amateur to distinguish between them and an expert can do so only after a microscopic study of imperfections. Ironically, the synthetic stones usually contain fewer flaws than the natural. Millions of carats of synthetic rubies are produced each year, but there is still a ready market for natural stones although they are a thousand times more expensive than the synthetic.

Until 1947, no star ruby or star sapphire had been synthesized; then the Linde Company in the United States announced the manufacture of "stars." One can now buy a synthetic star ruby or star sapphire not only practically indistinguishable from the natural stone but rivaling it in beauty. Several other minerals have also been synthesized by the Verneuil process, the most important among them being spinel, a magnesium-aluminum silicate. Boules of this material can be colored by adding metallic oxides. With vanadium as a pigmenting agent, a spinel can be made that resembles alexandrite, the gem variety of chrysoberyl that is red in artificial light and green in daylight. The prospective buyer should be aware that in certain places these synthetic stones are sold as genuine alexandrites.

Corundum in Industry

Corundum also serves as an industrial mineral. Because of its extreme hardness, exceeded only by the diamond, it is the most important natural abrasive material. Since early Egyptian times a black magnetic material, emery, has been used to cut, grind and polish hard rock and ornamental stone. Its chief source is Naxos, an island in the Grecian archipelago where mining has gone on for thousands of years.

Emery was thought to be a single mineral species until a study in the middle of the eighteenth century showed it to be a fine-grained mixture of two minerals: magnetite and corundum. The magnetite contributed the magnetic properties and the corundum the abrasive quality. Following this discovery, corundum was widely mined, crushed and manufactured into abrasive disks and wheels. At the beginning of the twentieth century, corundum was produced by the ton at a relatively low cost, using bauxite, the abundant ore of aluminum, as the raw material and electricity as the heat source. Together with the manufacture of other abrasive materials, this ended the mining of corundum as an abrasive. Nevertheless, there is still some demand for natural emery and the ancient mines of Naxos are therefore still operating.

Emeralds Yesterday and Today

The mineral beryl parallels corundum in that gem varieties of it are based on color. Commonly found in pegmatites, its colors range from colorless to pale green, yellow (known as golden beryl), blue-green (known as aquamarine), pink (known as morganite) and shades of other colors. Emerald, the deep green variety of beryl, is not found in pegmatites but in other geologic environments. Aside from its color, resulting from minute amounts of chromium or vanadium, emerald is the same as other beryl gems of lesser value.

The most ancient emerald mines, in Egypt's Zabara Mountains on the coast of the Red Sea, date back to the Twelfth Dynasty, about 1900 B.C. Much later, small emeralds were mined in the Salzburg Alps and it is believed that the Romans obtained stones there.

In 1830 emeralds were discovered in the Ural Mountains near Ekaterinburg (now Sverdlovsk) when a green stone in the roots of an overturned tree caught the eye of a peasant charcoal burner. As in Egypt, the crystals are embedded in mica schists. Although many fine stones have come from this locality, the larger ones are badly flawed and not of gem quality. Emeralds have been recovered in small quantities from the Transvaal, South Africa; Ajmer, India and Habachtal, Austria. The most recent find, and one that may prove to be significant, is in a remote part of the Mweza Range in Sandawana, Rhodesia. The stones are small but superb in color and quality.

By far the main world source of emeralds is Colombia, South America, with the most important mine located at Muzo, about ninety-four miles from Bogotá. When the Spaniards arrived, the Indians of Colombia, Ecuador and Peru already had a great store of emeralds, many of large size. The conquerors took these stones but the Indians refused, despite torture, to disclose the source of their treasure. The Cosquez mine was located quite by chance in 1558. According to legend, in 1594 the Muzo mine was also found by accident. The story tells how a Spanish horseman discovered that an emerald embedded in a hoof was causing his horse to limp. On retracing his course he found the Muzo mine, which, since that time, has been the world's principal producer of emeralds.

The Colombian emeralds are found in "pockets" in horizontal veins of calcite traversing a black carbonaceous limestone. Associated with them are crystals of quartz, some clear, others green, along with well-formed crystals of pyrite. The emeralds are typically hexagonal prisms rarely exceeding two inches in length, and most are much smaller. They are usually cracked and flawed, and larger crystals frequently fall to pieces shortly after being mined. Flaws are characteristic of emeralds and thus large flawless stones are great rarities and are among the most valuable of all gems.

At the time of the Spanish conquest, the presence of Colombian emeralds in Peru, Ecuador and even among the Mayas in Yucatan pointed to a very widespread trade among the Indians in northern South America and Central America. Archeologists question whether this trade reached Mexico City, but a remarkable specimen in the Vienna Museum argues that it did. The known history of this specimen, obviously of Colombian origin, begins in 1520. Among the treasures rested from Montezuma by Cortes was a foot-high piece of limestone of roughly conical shape. Partially embedded in it were ten deep-green emerald crystals varying in size from one to two inches long and from one half to one inch wide. With other treasures from the New World this specimen was acquired by Charles V of Spain, and it was handed down from one Hapsburg to another until it finally came into the possession of Francis

Joseph of Austria. In 1900 Francis Joseph turned the emerald over to the newly founded museum in Vienna. To keep it from falling into the hands of the Nazis during World War II, it was stored in the salt mines at Salzburg. After the war it was returned to Vienna but during the Russian occupation it was hidden beneath a rubble of bricks and mortar in a bombed-out room of the museum. Still one of the world's finest mineral specimens, it is on exhibit at the Vienna Museum today.

As with other precious stones, man has long dreamed of making emeralds. Their synthesis was reported in 1848 and again in 1885 by crystallization from a melt, but the emeralds were too small to be considered gems. Stones large enough for cutting were made by the German chemical trust, I.G. Farben, in the 1930's by an undisclosed method, but their production was apparently ended by World War II. Not until 1946 did synthetic emeralds enter the gem market. Carroll F. Chatham of San Francisco had for years experimented tirelessly with compounds of beryllium, aluminum and silicon, the basic elements in emeralds, before he finally produced a stone of sufficient size and quality for cutting. Since that time "Chatham emeralds" have been used in increasing numbers in fine jewelry throughout the world. Although emeralds have since been synthesized by others using known methods, Chatham has kept his process secret. The crystals he grows are not only identical to the natural mineral in hardness, luster and color, but have the same hexagonal crystal form. Unlike a synthetic ruby, which can be produced in a few hours, the growth of a ten-carat synthetic emerald takes many months. A much longer time must have been required for Chatham to produce the 1014-carat emerald crystal on display in the Smithsonian Institution. The green color of synthetic emeralds results from the introduction of chromium, the same element that is present in small quantities in most natural stones. Small amounts of other elements yield a variety of colors: nickel renders the stones olive-green, manganese produces a pale-pink, and cobalt a deep rose-pink.

With advancing technology new methods will undoubtedly be developed for the manufacture of inexpensive synthetic emeralds. It is unlikely, however, that these will threaten the value of natural stones for, as with rubies, certain buyers will pay a high price simply for the pleasure of owning a natural product.

Emeralds in the form of rough crystals and cut stones from Colombia. (Studio Hartmann)

15 Quartz

(Above) Quartz studded with crystals of green tourmaline from Minas Gerais, Brazil.

(Below) Needle-like crystals of rutile from Minas Gerais, Brazil, oriented by the structure of the colorless quartz in which they are enclosed, form a six-pointed star. (Both by Studio Hartmann)

Of the nearly two thousand mineral species none compares with quartz in diversity of origin and occurrence, in abundance of varieties, and manifold uses. As a major rock-forming mineral, it crystallizes at high temperatures from molten magmas; and yet it can be deposited on the sea floor from water at temperatures only a few degrees above freezing. It is widely sought and highly prized as a gem stone, and at the same time it is separated from ores to be discarded as worthless. As the major constituent of sand it is sold for only a few dollars a ton for lowly but important uses, such as the aggregate in concrete and for sanding slippery highways. At the other extreme, quartz brings many dollars a pound as water-clear crystals to be cut for technical purposes. From the flints of the stone age to the oscillator plate of the electronic engineer, quartz has proved to be one of the most important minerals used by man to shape the world in which he lives.

No other material played so large a part as flint in the story of man's first efforts to master his hostile environment. The story of flint is told elsewhere in this book, but that is only the opening scene in the drama of man's dependence on quartz through the ages. To our remote ancestors, the hunters of the Old Stone Age, failure of the flint supply might have meant death by starvation; today, the complete withdrawal of quartz as a raw material would mean that we could make neither glass nor concrete, and numerous less important aspects of our technological culture would be crippled. Keeping in mind the importance of quartz through the ages, let us examine some of the less well-known ways in which it has been of use to man and to the mineralogist in the development of his science.

Before the advent of writing and record-keeping about 3500 B.C., engraved seals were used to indicate possession and identity. Thus, a merchant shipping goods by caravan would close the necks of jars and packages with lumps of clay impressed with his personal seal. Kings signed their decrees with such seals, and with them priests indicated the approval of the gods of earthly affairs. These seals were either flat, like a modern seal ring, or cylindrical, with a hole passing through the long axis, or conical, and were made of baked clay, stone or hard mineral. The earliest seals, dated to 6500–6000 B.C., were made of relatively soft materials that could be engraved with tools of bird bone; but by 3000 B.C., as craftsmen learned to engrave harder materials with rotating bow drills and abrasive powders, hard minerals such as quartz became popular.

One inscription on a Babylonian cylinder seal of about 2000 B.C. reads (as translated by E. A. W. Budge): "A seal of Du-Shi-A (rock crystal) will extend the possessions of a man and its name is auspicious." Another reads: "With a seal of Gug (red carnelian or jasper) a man will never be separated from the protection of his god." These indicate the blend of mysticism with utility that often characterized the attitude of ancient peoples toward their mineral raw materials. Quartz is hard, homogeneous, of attractive appearance, and can be cut with a rotating drill charged with fine sand and water; hence, it is an ideal material for cylinder seals. At the same time, the ancient craftsman made an effort to choose a material that was in tune with the mysterious forces that dominated his world. Lest the reader wonder at the ability of ancient craftsmen to work with such hard materials, we may remark that we have seen modern Florentine craftsmen cutting agate and similar hard stones for mosaics with bow saws, bow drills, abrasive and water, much as their Sumerian counterparts must have done five thousand years ago.

Theophrastus, friend and pupil of Aristotle, writing about 300 B.C., in his treatise *On Stones* mentions *krystallos,* or rock crystal, *amethyst* and *sardion,* a red

When the outer surface of a rock cavity in Brazil, filled with colored layers of agate deposited one on the other, was ground away, the intercepted agate layers appeared as concentric circles in this eye agate.
(Harry Groom)

A delicate group of transparent quartz crystals from Dauphiné, France. (Emil Javorsky)

chalcedony. ". . . among the ancients, he writes, there was no precious stone in more common use; *onychion,* onyx, or banded chalcedony; *achates,* or agate, which he tells us came from the River Achates in Sicily; *prasitis,* possibly our heliotrope or bloodstone; and some other varieties of disputed identity. In this, the earliest work on minerals that has survived, quartz and its varieties is mentioned more often than any other mineral, and great value is placed on it as the stone from which seals were cut.

Pliny, in the first century A.D., wrote of quartz in his *Natural History* that it was "not easy to find out why nature should build with six-angled bricks," a noting of the dependence of external form on internal structure more than fifteen hundred years before the recognition of such a relation.

Quartz and Mineralogical Theory

More than any other mineral, quartz has been the birthplace and the testing-ground for the concepts and methods of mineralogy. It was on crystals of quartz that in 1669 Nicolaus Steno first observed the principle of the constancy of interfacial angles. Robert Boyle, in 1672, in *An Essay on the Origin and Virtues of Gems* exploded the two-thousand-year-old legend that quartz crystal was ice frozen so hard that it could not be melted. He measured the specific gravity of quartz and demonstrated that it was more than two and a half times too heavy to be water in any form. By his observations of liquid inclusions in quartz crystals, he correctly concluded that they were formed by crystallization from watery solutions.

Huyghens, in 1678, in his pioneer *Treatise on Light* described the double refraction of quartz as follows: "The double emission of waves of light, which I had imagined, became more probable to me after I had observed a certain phenomenon in the ordinary Rock Crystal. . . . For having had cut from it some well-polished prisms of different sections, I remarked in all, in viewing through them the flame of a candle or the lead of window panes, that every thing appeared double, though with images not very distant from one another." In a later paragraph, he adds: "Rock Crystal grows ordinarily in hexagonal bars. . . . It seems that in general the regularity which occurs in these productions comes from the arrangement of the small invisible equal particles of which they are composed." This clear and reasoned suggestion of internal structure as the cause of external form and physical properties antedated the final proof by nearly 350 years.

In 1772, a Frenchman, Rome de l'Isle, made precise measurements of the angles of quartz crystals and confirmed Steno's opinion that these angles do not vary with changing size and shape of the chrystal. The "Father of Crystallography," René Just Haüy, saw correctly that the pyramid-like terminations really consisted of two sets of three faces, at both the top and bottom ends of the crystals (r and z faces, Fig.1, page 235). Even more interesting was the observation that quartz crystals were of two kinds that corresponded to each other as the right hand does to the left. On many quartz crystals, Haüy found, there were small sloping faces either on one's left or right, as one faces the crystal.

Milky quartz crystals from Ouray, Colorado, are encrusted with smaller crystals of colorless quartz.
(Wilbert Draisin)

An irregular pile of dolomite crystals stands beside a beautifully formed transparent quartz crystal from Switzerland. (Alfred Ehrhardt)

In the century and a half since then it has been found that nature affords just about equal numbers of right-hand and left-hand quartz crystals, and that all quartz crystals are one or the other.

In 1814, Biot showed that quartz caused the plane of polarization of light to be rotated either to the right or the left. And in 1821, Sir John Herschel observed that those quartz crystals that had left-handed beveling faces rotated the plane of polarization to the left, whereas right-handed crystals rotated the plane of polarization in the opposite direction. This rotary polarization of light is observed in many crystals but none shows it as well as quartz.

Although the chemical composition of quartz was unknown in the eighteenth century, the great Swedish chemist Bergmann showed that it was identical with the "earth" called silica and was possibly an elemental substance. It remained for his even more famous countryman, Jons Jakob Berzelius, to break down the "earth" silica into oxygen and the new element, silicon. Thus, in addition to its other distinctions, quartz has contributed an element to the periodic table.

In 1880, Jacques Curie and his brother Pierre, later to gain fame as the co-discoverers of radium, were working on the electrical conductivity of crystalline bodies at the University of Paris. In attempting to measure the electrical conductivity of plates of quartz, the brothers observed that pressure on the test plates produces a deflection of the sensitive electrometer. They had discovered the property of matter known as piezoelectricity (from the Greek *piezon*, meaning pressure). The piezoelectric property in quartz is today the source of a multimillion dollar industry. It was found in 1921 that oriented slices of quartz, properly mounted, would vibrate mechanically at radio frequencies, and hence control, or stabilize, the frequency of a radio transmitter. The frequency of transmitters today are controlled by a vibrating slab of quartz. Thus the fact that you find a station at its accustomed point on the dial derives from a little wafer of quartz and the keen observation of two young Frenchmen.

The structure of quartz is rather complex, in spite of its chemical simplicity, but was finally worked out in 1926 by the English physicist, Reginald E. Gibbs.

Opal as the petrifying agent has so faithfully replaced the woody structure in this specimen from New Mexico that a botanist can identify the tree from which it came. (Emil Javorsky)

(Right) Crystalline crusts of colorless to amethystine quartz surround cores of multicolored agate. (Studio Hartmann)

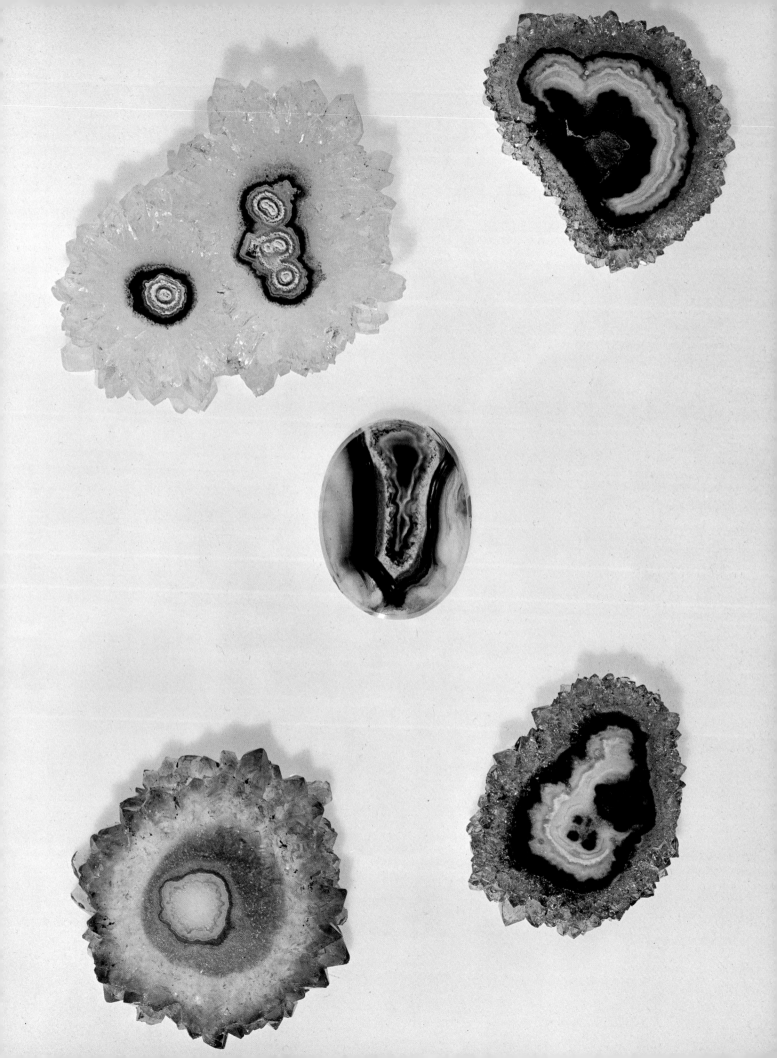

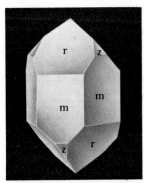

Figure 1

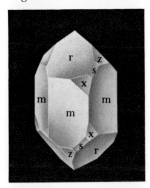

Figure 2

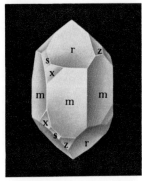

Figure 3
Quartz Crystals
(Left above) A polished slice of agate from Uruguay composed of concentric bands of vari-colored chalcedony with a central core of coarsely crystalline quartz.
(Left) Translucent chalcedony from South Africa shows its characteristic waxy luster.
(Both by Studio Hartmann)

He found that the silicon and oxygen atoms are so joined that each silicon is surrounded by four oxygens, as at the corners of a tetrahedron, and that each oxygen is shared by two silicons. The tetrahedral groups are arranged like a spiral staircase about the long axis of the quartz crystal. If the staircase spirals to the left, it is a left-handed crystal and will rotate the plane of polarization to the left. The same applies to the right-handed crystals, with the opposite sense throughout. The questions raised by Pliny, Huyghens, Haüy and other investigators through the centuries were answered at last. The ultimate structure of quartz was known.

Knowledge of the generalized structure does not answer all questions about a mineral, because not all crystals of a species are perfect; in fact, perfect crystals are unknown either in nature or technology. Variations in composition and in conditions of crystallization produce a corresponding variation in properties, and hence in interest to man. It is, therefore, as necessary for us as for Theophrastus and Pliny to describe natural occurrences, varieties, and limits of variation in properties. We must also do what they could not do: relate these properties to our knowledge of crystal structure and chemical composition.

Quartz is silicon dioxide, SiO_2, in which silicon and oxygen in ratio 1:2 make up the tiny building units of its internal structure. The arrangement of these units determines whether the mineral is quartz or another mineral with the same chemical composition. At present we know of ten of these silica minerals, each with different properties resulting from different arrangements of identical units. Compared with quartz, all the others are of minor importance. Quartz occurs in well-formed crystals more frequently than any other mineral. Crystals most commonly have a hexagonal outline with six elongated prism faces terminated by six sloping faces (Fig. 1), in which the *r* faces are larger than the alternating *z* faces. The rare *s* and *x* faces are those which, by their position indicate the "hand" of a crystal, right (Fig. 2) or left (Fig. 3).

This characteristic habit makes identification of crystals easy. However, quartz is characterized by other physical properties that enable one to determine it even when not in crystals. It is hard ($H = 7$) and will easily scratch glass. Its specific gravity is 2.65, which is considered an average for nonmetallic minerals. Furthermore, unlike most minerals with which it occurs or with which it otherwise might be confused, it has no cleavage, but breaks with a conchoidal fracture in a manner similar to glass. Most crystallized quartz is white or colorless, but also is found in several different colors, each of which is given a varietal name. Although traces of impurities are found, chemical analyses show that quartz closely approaches the theoretical composition of silicon dioxide.

The many varieties of quartz are divided into two groups on the basis of texture—that is, coarse-crystalline or fine-crystalline. Each of these groups is subdivided principally on the basis of color. The color may be caused by chemical impurities incorporated within the crystal structure or it may be the result of the mechanical inclusion of other finely divided mineral particles. Of the host of names given through the centuries to these varieties only a few are in common use today. Although German miners had for centuries used the name quartz in a restricted sense, it was not until 1783 that Rome de l'Isle used the name in its present all-inclusive sense.

Coarse-Crystalline Varieties of Quartz

The coarse-crystalline varieties include clear, colorless or milky white quartz, purple amethyst, yellow citrine, dark smoky quartz, and pink rose quartz. They all have the same crystal structure and the same external crystal forms; it is only in color that they differ.

Well-formed crystals of clear, colorless quartz have long gone under the name of rock crystal, but for many centuries earlier they were called merely crystal. This word, now applied to the ordered state of matter and to the resulting external shapes, was taken from the Greek *cristallos,* meaning ice. No matter what other conditions exist at the time of formation, a mineral must be free to grow into an open space if it is to form its characteristic crystal shape bounded by smooth plane surfaces. Thus, rock crystal is found in open veins and rock cavities in which growth ceased before the enlarging crystals interfered with each other.

Crystals of clear quartz have been found in many parts of the world in sizes ranging from the microscopic to one weighing $5\frac{1}{2}$ tons. This largest crystal came from Minas Gerais, the state in Brazil that supplied most of the rock crystal during the twentieth century.

Large flawless quartz crystals have been in demand since before Christ as the raw material from which "crystal" vases, goblets, bowls and other objects, both functional and artistic, were carved. With the development of high-quality glass, objects have been made from glass that resemble those formerly made from rock crystal. The term "crystal" is retained for them; thus today one can buy "crystal" goblets or a "crystal" vase, and the fortune-telling medium gazes into a "crystal" ball. All these are made from glass, nearly the only solid material that is noncrystalline according to our present definition.

When China was opened to world trade, it was discovered that the art of carving natural materials was already highly developed there. The skill of Chinese craftsmen in carvings in rock crystals and other minerals is now known everywhere. Much of the raw material did not originate in China but was sent there for the relatively inexpensive work of carving. The Chinese are also skillful in cutting and polishing spheres from flawless quartz and these true crystal balls can be seen in many mineral collections. The outstanding crystal ball, on exhibit in the United States National Museum, is a perfect sphere with a diameter of nearly thirteen inches and a weight of 107 pounds.

Far more abundant than rock crystal are other colorless or milky white types of coarse-crystalline quartz. It is the most common mineral in veins where it may occur alone or in association with ore minerals. Nearly pure quartz masses weighing thousands of tons form a major portion of many pegmatites. Although less conspicuous, the chief occurrence of quartz is as a rock-making mineral disseminated in small grains through thousands of cubic miles of the earth's crust. As rocks and veins weather, the quartz is liberated from its matrix, broken into small grains and washed as sand into neighboring streams. When it reaches the sea, currents may sweep it along the coast to form the sand of the seashore, or carry it into deeper water where it settles to the sea floor and builds up thick deposits. Accumulations from a few feet to thousands of feet thick have formed in many parts of the world. When the sand grains become firmly cemented together, the deposit becomes a sandstone.

Polished slices of agate from Morocco. (Studio Hartmann)

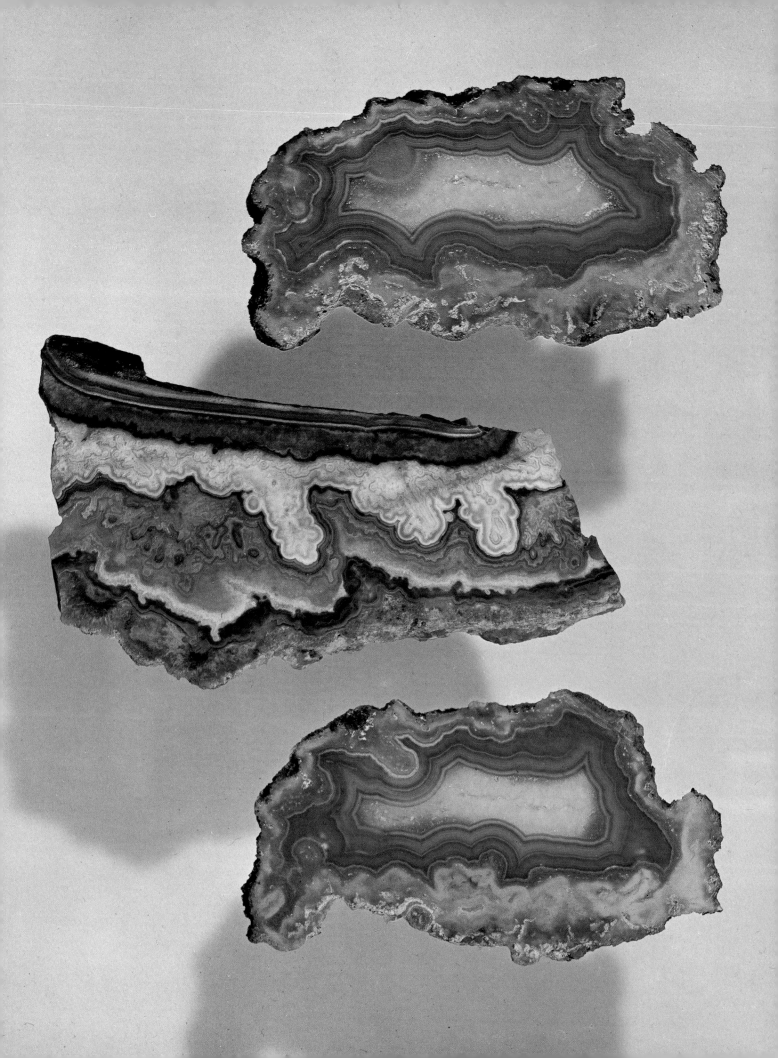

Amethyst

Amethyst as an ornamental stone was known and valued in antiquity. It is mentioned in the Bible as one of the twelve gem stones in the breastplate of the High Priest and was inscribed with the name of one of the twelve tribes of Israel. Amethyst was believed to endow the wearer with many supernatural qualities such as quickening his intelligence and making him invulnerable in battle. But its special virtue was the capacity to cure or prevent drunkenness. This belief was so strong that the name amethyst is derived from a Greek word meaning "not drunken." Aside from these supposed virtues, amethyst has been prized for its intrinsic beauty, and with the advancing skill of the lapidary was cut into valuable gem stones. But, as with any gem, the value is dependent on supply and demand. Thus, when large amounts of amethyst were discovered in Brazil and Uruguay at the beginning of the nineteenth century, the price fell rapidly.

Although amethyst is distinguished because of its violet color, it varies from pale shades to a deep rich purple, the value increasing with the depth of color. In all shades one can discover on close examination that the color is not uniformly distributed through the stone but is concentrated in thin sheets interlaminated with colorless sheets. This color layering, which is parallel to the r and z crystal faces (Fig. 1, page 235), is so characteristic of amethyst that its absence should lead one to suspect an imitation.

Chemical analyses of amethyst show it to be nearly pure SiO_2 but always containing some iron (less than 0.10 per cent). Since the amount of iron increases with increasing depth of color it is assumed that this element is the coloring agent. Although the color is completely stable at ordinary temperatures, heating produces a remarkable change. If it is heated to 450° C the mineral becomes colorless; but if heated to 550° C it turns to a yellowish-brown, the color of citrine. Much of the citrine sold as gem stones is probably a poor grade of amethyst heat-treated in this manner.

The quartz that occurs as citrine in nature is rarer than amethyst but is found in the same localities and formed under the same conditions. Its yellow to yellow-brown color, resembling closely that of gem topaz, has given rise to a confusion in the gem trade. Although topaz is by far the more valuable gem, in some countries of the world citrine is sold to the unwary buyer as "genuine topaz." Elsewhere, as in the United States, it is commonly referred to as "quartz topaz" or "topaz quartz" to distinguish it from the much more costly "precious topaz."

Smoky Quartz and Others

Smoky quartz, as the name implies, has a smoky color that varies in intensity from a pale-brown to nearly black. As in amethyst the color is rarely uniform but is concentrated in bands parallel to crystal faces and, as in amethyst, interesting effects are produced by heating. If a smoky quartz crystal is heated at 450° C the "smoke" vanishes in a few minutes, leaving the crystal colorless, but unaffected in other ways. The color can be restored by irradiating the bleached crystal in a beam of x-rays. For this reason and because chemical analysis fails to detect any impurity not present in colorless quartz, many

The great range of color and patterns in agate is illustrated in polished slices from Mexico. (Cut and polished by Joe Guetterman. Photo by John H. Gerard)

believe that the color of smoky quartz is the result of a natural irradiation.

Rose quartz, unlike the other coarse-grained varieties, is rarely clear or in well-formed crystals but is usually massive and has a murky appearance. The color that ranges from a barely perceptible shade of pink to a deep rose-red is attributed to small amounts of titanium within the quartz structure. Further, the mineral rutile, titanium dioxide, is frequently present in rose quartz as microscopic needles oriented along crystallographic directions. In a sphere cut from such material the light is scattered by the oriented inclusions forming a six-pointed star on the surface.

Rose quartz is found almost exclusively in pegmatites, frequently in large masses crossed by veinlets of milky-quartz. In some places, as in the Black Hills of South Dakota, it is quarried for decorative purposes. The rare material free from cracks and with a deep rose-red color is used for jewelry and carvings.

So rare are well-formed crystals of rose quartz that a textbook written in 1940 declared they did not exist. Since that time small crystals have been found in Maine, notably at Newry in Oxford County. Much more recently, larger crystals in beautiful aggregates were discovered in a pegmatite near Governador Valadoris, Minas Gerais, Brazil.

Transparent quartz crystals frequently incorporate other minerals which may impart various colors to them. Thus, crystals are colored green by the presence of chlorite and red from included hematite. If these foreign minerals are deposited on the faces of the transparent quartz only during brief stages of growth, the outline that one sees within the crystal is called a phantom or ghost crystal.

Colorless quartz crystals shot through with long slender needles of rutile are of particular interest. The individual needles in this rutilated quartz (also called Venus' hair stone) are coarse enough to be easily seen and they thus impart no general color to the enclosing quartz. Rutilated quartz is today used in inexpensive jewelry but during the eighteenth century polished specimens were highly valued as ornamental objects in England and France.

Fine-Grained Varieties of Quartz

The varieties of quartz included under this heading all lack any external evidence of crystallinity. Yet their chemical composition and crystal structure show that they are indeed quartz. The tiny crystalline particles of which they are composed are so intimately intergrown as to give a homogenous texture to the aggregate. The two kinds of intergrowth, fibrous and granular, usually can be distinguished only with a microscope. There is no special designation of the granular types but the fibrous type goes under the general name of chalcedony of which there are many subvarieties, based mostly on color.

The term chalcedony is used in a more restricted sense to specify a certain subvariety which occurs as crusts with a rounded hummocky surface partially filling open spaces as gas cavities in basaltic lava flows. It is translucent with a waxy luster and colored a pale shade of blue, yellow, gray or red. The variously colored subvarieties of chalcedony can be cut and polished to a high luster and have been used from early times in jewelry, inlay work and ornamental carvings. Although subtle differences in color have given rise to many names, the following are most important:

Diatoms are microscopic single-celled algae living in either fresh or seawater. Their shells, some illustrated here greatly magnified, are opaline silica secreted from the water. When the organisms die, their shells settle to the floor and in places accumulate in great thicknesses as the rock, diatomite. (Johns-Mansville Photo)

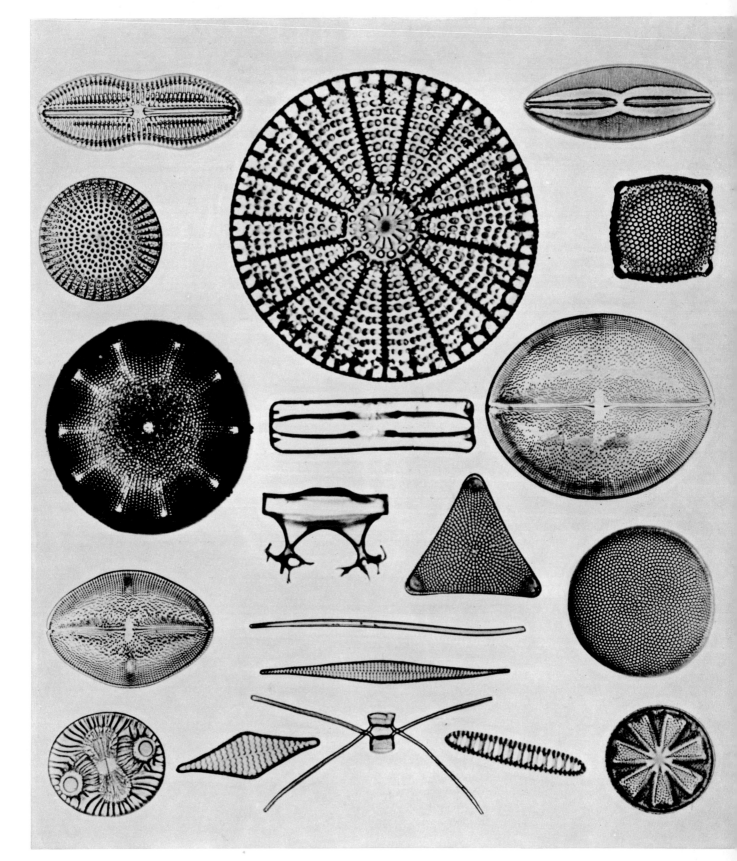

In 1866, W. H. Barton used this carefully shaped slab of diatomite as a claim marker in the Silver Peak District of California. (Cornelius S. Hurlbut, Jr.)

Carnelian and sard are translucent types of chalcedony with no well-defined distinction between them. The color, resulting from finely dispersed iron oxide, ranges from a deep-red in the most highly prized carnelian through reddish-brown to chestnut-brown in sard. The red stones so beautifully polished and exquisitely shaped in the inlay work of the Taj Mahal are carnelian.

Agate

Of the several varieties of chalcedony, agate is most common and thus best known and most widely used. It occurs usually as cavity fillings with layers or bands differing in color and transparency and alternating with one another. The individual layers, sometimes delicately fine but with uniform thickness and color, are arranged concentrically parallel to the walls of the cavity. If

(Right) Colors resulting from internal reflections give this Australian gem opal in the matrix its subtle beauty. The four small yellow-to-red brilliant opals are from Mexico. (Studio Hartmann)

this surface is angular and irregular the layers in cross section may show mimetic patterns to which names such as landscape agate, fortification agate, star agate, and brecciated agate are given. The natural colors of the bands vary from white through gray to black but may be pale shades of red and brown, and less commonly blue, green or lavender.

Although the crystal structure of agate is identical to that of coarsely-crystalline quartz, the specific gravity is slightly less, indicating a porosity. That this porosity is not uniform is shown by the fact that some layers will absorb and be colored by a dye whereas others remain unchanged. Most of the agate that is used in jewelry and for ornamental purposes has been artificially colored by soaking in various chemical dyes or pigmenting solutions. The interesting process used to color agate black has been known for a long time. The cut pieces are immersed in a warm solution of honey or sugar for a sufficiently long time to allow the solution to diffuse into the porous layers. They are then removed, cleaned, and immersed in sulfuric acid which carbonizes the sugar rendering the agate black.

Onyx is a variety of agate in which the alternating bands instead of being curved or concentric are plane and parallel. It is onyx composed of black and white layers that is used for cameos, with the carved figure in the white layer standing in relief on a background of the black layer.

Moss agate is the name given to a white or gray translucent chalcedony in which darker included material has a form resembling moss or other plant life. The mosslike inclusions are usually black and more rarely red or brown. The black color is believed due to manganese oxide and the red and brown to iron oxides.

Chrysoprase, plasma and prase are green varieties of fine-grained quartz, sometimes of the fibrous chalcedonic type and sometimes of the granular type. Chrysoprase is translucent with a bright-green color resulting from the presence of a hydrated nickel silicate mineral. It is thus commonly associated with nickel deposits, particularly those found in serpentine rocks. Plasma is a very dark-green, almost opaque, mineral with the color resulting from disseminated particles of green minerals such as chlorite and amphibole. Bloodstone or heliotrope is a type of plasma through which are scattered dark-red spots resembling blood. Prase has a leek-green color resulting from the inclusion of many tiny fibers of an amphibole. It is more translucent than plasma and one can frequently see the individual pigmenting fibers. Because of the similarity to jade in color, prase is widely used as an inexpensive imitation of jade.

Flint and chert, granular types of fine-grained quartz, have a similar occurrence and origin and no sharp distinction exists between them. They break with a conchoidal fracture yielding sharp edges and they were used by early man as cutting tools and weapons. Although they are both found in sedimentary rocks, the term flint is usually reserved for isolated nodules found in chalk or marly limestones. Chert also is characteristic of limestones but occurs as bedded layers ranging from a few inches to a few feet thick, and frequently of great lateral extent. Moreover, a color distinction can usually be made: flint is dark-gray to black whereas chert is white to light-gray.

Jasper is a granular type of fine-grained quartz usually colored red by the inclusion of finely divided hematite, but may be yellow or brown by the inclusion of other iron oxides. Its occurrence is much like chert forming in

(Above) Banded gem opal in brown limonite from Australia. (Benjamin M. Shaub)

(Below left) Rough and polished fragments of opal from Lightning Ridge, New South Wales, Australia. (Janet Finch)

(Below right) Opal in matrix from Mexico, and polished stones from Australia. (Rough stones from Washington University Collection; polished pieces from Walter C. Blatt, St. Louis, and William V. Schmidt Co., New York. Photo by John H. Gerard)

places thick beds with a wide areal extent. Jasper is the principal type of silica found in petrified wood although chalcedonic material and opal are also common. For the process of petrification to take place the wood must be buried. A single log can be covered by the sands of a stream or a whole forest can be overwhelmed and buried beneath a great shower of ash from a neighboring volcano. In either case, the wood eventually becomes saturated with water that slowly circulates beneath the surface. If the water contains silica in solution, as it will in an area of recent volcanic ash, silica in one of its fine-grained forms may substitute for the wood cell by cell until the log or tree has been completely replaced. In many places petrified wood can be found in which the external form and the cellular structure have been so faithfully preserved that it is possible not only to count the tree rings but also to determine the species.

Gem Opal

Gem opal or precious opal has always been valued for a subtle beauty resulting from an internal display of colors that makes it unique among gem stones. Various names have been given to precious opal such as fire opal, a stone in which red to orange colors are emitted from a translucent background, and black opal, which has a dark body color giving forth a brilliant play of color. One can scarcely improve on Pliny's description written nearly two thousand years ago. He states, "There is in them a softer fire than in the carbuncle, there is the brilliant purple of the amethyst; there is the sea-green of the emerald—all shining together in incredible union. Some by their refulgent splendor rival the colors of the painters, others the flame of burning sulfur or fire quickened by oil."

The great value placed on opal is conveyed by Pliny in relating the story of the Roman senator, Nonius, who owned a ring set with an opal the size of a hazelnut. The stone was coveted by the emperor, Marcus Antonius, who demanded it; but rather than yield, Nonius chose to abandon all his other possessions and flee the country.

The great value of opals, frequently equal to that of diamonds, continued into the nineteenth century. The decrease in popularity resulting in a lessening of value is attributed in part to the superstition that opals were unlucky stones. Many believe that this foolish notion was initiated by Sir Walter Scott in *Ann of Geierstein*, published in 1829. The tale revolves around an enchanted princess, Lady Hermione, who always wore a lustrous opal in her hair. The stone also was enchanted and reflected the moods of its wearer, glowing when she was gay, but emitting fire-red flashes when she was angry. When by accident a few drops of holy water were sprinkled on it, the brilliance faded and Lady Hermione slowly sank to the chapel floor. She was placed on the bed in her chamber but the next morning only ashes remained where she had lain. Scott might well have had Lady Hermione wear a diamond, ruby or emerald but probably chose the opal because it changes in color and luster pending on the light and the position of the observer. It is reported that within a year after the publication of *Ann of Geierstein* opals depreciated in value at least fifty per cent.

The finest gem opals of the ancients are believed to have come from India but stones of lesser quality were also found in Arabia and Egypt. From Roman

Tiny diatoms accumulated on the sea floor and built up a thickness of four thousand feet of diatomite in the white cliffs at Lompoc, California, exposed during mining operations. (Johns-Mansville Photo)

times, and continuing until the discovery of opals in Australia, Hungary was the principal source of high-quality precious opal.

Of the several important Australian opal localities, the first were located in Queensland in 1872. These were followed by discoveries in New South Wales of the White Cliffs area (1899) and the Lightning Ridge field (1908). Several later finds were made in South Australia and Western Australia. Although all localities produce precious opal of fine quality, the stones from each district have characteristics by which they can usually be recognized. The Queensland opal has a milky white body color with a blue to green play of colors whereas black opals are produced from Lightning Ridge. At White Cliffs, gem opal, replacing crystals of the mineral glauberite, has the shape of fossilized pineapples and animal bones; shells and wood have also been converted to masses of precious opal.

Several important opal localities are in the state of Queretaro, Mexico, where fire opals showing a wide range of color are found. In the United States the outstanding locality for precious opal is the Virgin Valley, Humboldt County, Nevada. Here wood embedded in volcanic ash has been partially or completely transformed into opal.

Although the greatest interest attaches to precious opal, common opal, of which there are several types, is far more abundant. Opal is a form of silica, but unlike quartz and the other varieties, contains variable amounts of water ranging from four to twenty per cent. Although early studies indicated it was amorphous, completely lacking crystal structure, recent x-ray analysis shows it to be essentially an aggregate of submicroscopic crystallites of cristobalite. Cristobalite is one of the crystalline varieties of silica. Because of the presence of water, opal has a lower hardness (5½–6½) and specific gravity (2.00–2.25) than quartz. Pure opal is colorless or milky white but frequently incorporates finely divided particles of other minerals that, acting as pigmenting material, impart to it almost every color. In addition, precious opal shows the play of colors resulting from the interference of light.

Opal is a low-pressure, low-temperature mineral and is formed at or near the earth's surface by circulating ground water or deposited from rising hydrothermal solutions as they approach the surface. It is thus found with minerals of the traprock suite in cavities and cracks in basaltic lava flows, and as petrifying material in recent volcanic ash and sedimentary rocks. A porous whitish-gray type of opal known as *geyserite* or siliceous sinter is deposited on the surface by hot springs and geysers as at Old Faithful in Yellowstone National Park.

Diatomite or diatomaceous earth is a porous chalky-like material that has been deposited as a sedimentary rock on a lake bottom or sea floor. Although in appearance it in no way resembles other types of opal, diatomite is in fact a form of opaline silica. Diatoms are microscopic plants that have the ability to secrete silica from the water in which they live and to form from it hard shells of opal. On their death these tiny shells settle to the bottom. It is in such accumulations that the great bulk of the earth's opal is concentrated. Although an individual diatom cannot be seen with the unaided eye, their countless billions of skeletons have in places accumulated to tremendous thicknesses; the world's largest known deposit near Lampoc, California, varies between three thousand and five thousand feet thick. Here it is mined on a large scale and one can see dazzling white cliffs rising vertically scores of feet above the quarry floor.

The uses of diatomite depend on the microscopic structure of the individual diatoms and its great microporosity which results in a very low aggregate specific gravity. Its chief use is for filtering and purification of liquids but it is important as a filler in paint and paper, a heat insulator, an absorbent for chemical disinfectants and dynamite, and an abrasive.

16 Colored Stones and Ornamental Minerals

Although many minerals have a dual nature, as for example, corundum, which may be used as an abrasive or be treasured as a priceless ruby, some are merely utilitarian and others have been cherished for their beauty alone. Members of this last group that have not been considered elsewhere are brought together here. Some are cut as gem stones while others are carved or cut into slabs and polished for decorative and ornamental purposes.

Lapis Lazuli

The mineral lazurite gets its name from its deep azure-blue color. The small crystals of it that are occasionally found occur in the form of a twelve-sided solid, the dodecahedron. But the interest in this mineral through the centuries has been not in single crystals nor the pure material but rather as the major constituent of lapis lazuli. Lapis, as it is sometimes called, is a mixture of several minerals embedded in a matrix of white calcite. In addition to lazurite, augite and hornblende may be present and, almost invariably, small crystals of pyrite. The appearance of the tiny specks of brilliant pyrite on surfaces of polished material have been likened to stars in a blue firmament. In general, the higher the percentage of lazurite, the deeper the blue color, although green or violet shades may occur. For a gem material, lapis lazuli is relatively soft (hardness 5½), and if worn in jewelry quickly loses its polish.

Lapis lazuli was treasured in Egypt in predynastic time, probably prior to 3400 B.C. Several references made to it in the Old Testament under the name of "sapphire" indicate that it was known to the Hebrews long before any records were kept.

Lapis lazuli has been found in several places, but the principal source through the ages has been Badakhshan, near the northeastern border of Afghanistan. Although the mines of the area were ancient when visited by Marco Polo in 1271, they are still active producers. It seems likely that the lapis carved into scarabs by ancient Egyptians and the "sapphire" of Mesopotamia came from there. The mines are located high on the precipitous wall of the Kokcha Valley in a black and white limestone. Like all lapis deposits, these are the result of contact metamorphism: hot solutions from an intruded igneous magma reacted with the limestone to form the minerals that make up lapis lazuli.

Near the southern shore of Lake Baikal, in Siberia, lapis lazuli, known as *Siberian lapis* or *Russian lapis,* and of extremely fine quality, is found at the contact of a granite and a white limestone. Half way around the world, at Ovalle in the Chilian Andes near the border of Argentina, is another deposit of lapis lazuli. It is of poorer quality than the Russian but occurs in a similar geologic setting.

In addition to jewelry, lapis has been used for the carving of small objects such as figurines, vases and bowls. During Czarist times it was used extensively by the Russians as a decorative material for mosaics and inlay work. At the Hermitage in Leningrad one can see the finest examples of this art—urns and vases veneered with lapis lazuli and other decorative minerals.

Until late in the eighteenth century the artists' pigment ultramarine, was made from the blue constituent of lapis lazuli. Today a synthetic lazurite, in powder form, is a most satisfactory substitute for natural ultramarine.

Although lapis lazuli is not in itself of great value, substitutes for it are fairly common. Blue glass and agate dyed blue have both been sold under the name of lapis lazuli. The imitations can be easily detected since both glass and agate are harder and have a higher specific gravity than lapis.

Sodalite, another blue mineral, is sometimes confused with lazurite of lapis lazuli. But a comparison of them will reveal that sodalite has a violet shade, not the rich blue of lazurite. Large masses of sodalite have been found in several places in association with nepheline syenite rocks. A major occurrence is at Bancroft, Ontario, where it was at one time quarried as a decorative stone. It was not used extensively because the lovely blue seen by daylight looks almost black under artificial light.

Turquoise

As a gem and ornamental material, turquoise has at least as long a history as lapis lazuli. Like lapis, it is valued chiefly for its color, which varies from green to blue but is best described as a robin's egg blue. Turquoise is an aluminum phosphate but owes its color to the fact that it contains some copper. It usually occurs as a fine-grained amorphous appearing mass with a smooth fracture; crystals are extremely rare. Unlike most gem materials, turquoise is somewhat porous and will absorb liquids, grease or perspiration and change color accordingly.

Early peoples undoubtedly discovered turquoise because it is a secondary mineral; that is, it is found at or near the surface as a result of the action of ground water on underlying bedrock. The most ancient turquoise mines are at Sarabit Elkhadem on the Sinai Peninsula and may have been worked as early as 3400 B.C. The Arabs today obtain some turquoise there but the extensive

(Above) Deep azure-blue lazurite, brass-yellow pyrite and a matrix of white calcite combine to form a gem material, lapis lazuli, from Iran. (Ted Boente collection. Photo by John H. Gerard)

(Below) Lapis lazuli, shown here in a polished slice and cut stones from Afghanistan, lacks the hardness and brilliance of many gem stones but its color alone has always made it valuable. (Studio Hartmann)

underground workings are credited to the Egyptians; for two thousand years the pharaohs worked the mines with slave labor. The carved turquoise in the bracelets of Queen Zer of Egypt's First Dynasty, unquestionably from the Sinai Peninsula, may well make these the world's oldest jewelry pieces.

The most famous turquoise deposits are at Nishapur, Iran. The mines there, still active, may well date back to the beginning of the Christian era. The name turquoise, which means "turkish," was probably given to the mineral because it was originally brought to Europe from Persia through Turkey.

Aside from the Persian deposits, the only sources of importance today are in China, Tibet and the southwestern United States. Turquoise was used and treasured by the American Indians long before the Spaniards arrived. Many turquoise beads have been found in Pueblo Bonito, dating from A.D. 900 to 1000, and turquoise pendants were found in the much earlier "Basket Maker Village" in Chaco Canyon. Turquoise itself has come from many deposits in the American Southwest, but the most famous is in the Los Cerillos Mountains in New Mexico. The evidence of the extraordinary extent of the operations here is in an open cut two hundred feet deep and three hundred feet wide. Around this great man-made crater are dumps of waste material covering twenty acres. Trees six hundred years old growing on the dumps testify to the antiquity of the mining activity. The Spaniards worked this mine for one hundred years until caving ended their operations in 1686.

Variscite

Variscite closely resembles turquoise. Both minerals are nontransparent aluminum phosphates formed near the surface under similar conditions. The color of variscite varies from emerald-green to pale-green, but when cut as a gem, it may resemble a greenish-turquoise. It was first described in 1830 and minor deposits were thereafter reported in Germany, Austria, Czechoslovakia and the Congo. However, the first deposit of gem quality was discovered in 1893 near Lewiston, Utah; but most of the gem material has come from a more extensive deposit at Fairfield, Utah. At Fairfield, a crushed and brecciated zone in limestone has served as a channel for mineralizing solutions from which variscite, other phosphate minerals and chalcedony have been deposited. The variscite is found in nodular masses of varying size, some as much as a foot in diameter.

Rhodonite

The relatively extreme hardness and pleasing rose-red color of rhodonite (manganese silicate) make it a most satisfactory material for beads and small carvings. It is, however, unsuitable for use as exterior facings on buildings for in contact with rain it soon alters to a black manganese oxide. If, when exposed at the earth's surface, the altering solutions penetrate along cracks, the resulting thin black veins form a pleasing contrast to the rose-red of the polished stones.

Rhodonite has been found at Franklin, New Jersey, in large well-formed crystals. Elsewhere it is usually fine-grained, and in this form is found in small quantities in many places, most notably at Campiglia Marittima, Italy; Långban, Sweden; Vittinge, Finland; and in the United States at Cummington,

(Above left) Carving of nephrite jade illustrates the skill and attention to detail of Chinese artisans. (Harry Groom)

(Above right) The bluish-green color of polished turquoise, as in this nugget from Iran, is so characteristic that "turquoise blue" is frequently used to describe a color. (Katherine H. Jensen)

(Below left) This turquoise Buddha carved in Japan is remarkable for the absence of any trace of matrix material in it. (Harry Groom)

(Below right) A hunting fetish of turquoise carved by the Zuni Indians of New Mexico. An arrow point of pink agate is tied to the back of the effigy. (Tad Nichols)

Manganite, an ore mineral of manganese, occurs in well-formed crystals at Ilfeld, Germany. (Wilbert Draisin)

Massachusetts. The outstanding source is near Sverdlovsk, Russia, on the eastern slope of the Ural Mountains, where a large deposit has long produced rhodonite useful for ornamental stones. It has been used for large carvings as well as small ones; thus the sarcophagus of Alexander II of Russia was hewn from a single block. The Russians are still masters of the art of working with rhodonite; the Soviet building at the 1939 World's Fair in New York had a gigantic wall map of the Soviet Union that had been made by inlaying mineral plates of different colors with the landmasses appropriately represented by purplish-red rhodonite.

The lovely rose-red of rhodochrosite, the manganese carbonate, is similar to the color of rhodonite, but unfortunately the mineral is too soft to be used for ornamental purposes. It is generally associated with silver ores and forms botryoidal masses with concentric layers delicately tinted in shades of pink and red, as at Capillitas, Catamarca, in Argentina. In a few places single crystals are found, the finest coming from the silver mines at Alma and Alicante, Colorado. The ordinary ores of manganese are the black oxides, pyrolusite and psilomelane, and only rarely does rhodochrosite occur as an ore mineral. At Butte, Montana, however, it occurs in veins in such large quantities that it has been mined as an ore of manganese.

Idocrase

Mount Vesuvius, near Naples, Italy, has been built up by many outpourings of lava. In making its way to the surface, the molten magma has punched through horizontal layers of limestone carrying upward blocks of engulfed rock. Before being ejected from the volcano some of the limestone fragments

(Above) The polished piece of rhodonite, with characteristic rose-red color, from the Ural Mountains, Siberia, is crossed by veins of black manganese oxide; the rough crystals are from Franklin, New Jersey. (Emil Javorsky)

(Below) Polished slice of blue-green variscite cut from a nodule found at Fairfield, Utah, is veined by other phosphate minerals. (Studio Hartmann)

reacted with the magma to form new minerals. These are chiefly calcium-aluminum silicates; the calcium was contributed by the limestone, and the aluminum and silicon by the igneous rock. Idocrase is a mineral characteristic of such reactions and it has been found at Mount Vesuvius in brown transparent crystals suitable for cutting into gem stones. Because of its occurrence on the volcano some mineralogists have called the mineral *vesuvianite*.

Although idocrase as a gem is almost exclusively confined to Italy, the mineral is found in many parts of the world usually as a contact metamorphic mineral formed by the reaction of a limestone with an igneous magma. The color, varying from place to place, is most commonly brown or green but may be yellow, blue or red. It sometimes resembles jade, and a bright-green, massive variety, known as *californite*, found in Tulase and Siskiyou Counties in California, is made into carvings and ornaments and sold as jade.

Jade

Two mineral groups, pyroxene and amphibole, abundantly represented in rocks by augite and hornblende respectively, include more glamorous varieties in jadeite (pyroxene) and nephrite (amphibole). Although different chemically, they have in common the properties of extreme toughness and generally a green color; together they go under the name of jade. Microscopic study shows nephrite to be a dense felted mass of short fibers whereas jadeite is composed of an aggregate of microscopic grains. Although jadeite is slightly harder, the interlocking fibers of nephrite give it a greater cohesive strength, so that it is almost impossible to break a specimen with a hammer. It is frequently difficult to distinguish one from the other, particularly in polished pieces where the textural difference is obscured. A rather subtle distinction may be made on the basis of luster: nephrite is oily and jadeite is vitreous.

Nephrite is mineralogically identical with tremolite, which may be colored green by the presence of iron. As the iron in tremolite increases, the color becomes deeper and tremolite grades into actinolite. Thus nephrite may be milky-white to dark-green depending on the iron content. In general, it has a more uniform color than jadeite, which is frequently mottled. The spinach-green jade speckled with darker spots is nephrite, whereas the emerald-green stone colored by chromium is jadeite. Although green is the prevailing color of jade, blue, brown, yellow, mauve, gray and black, resulting from impurities, are known.

Among the Chinese, jade has for many centuries been esteemed for its beauty, surpassing all other substances in value, and working it into delicate and elaborate pieces is an ancient art. Evidence supports the belief that the jade was not obtained locally but was brought by caravan from Chinese Turkestan. There, from stream beds in the mountains of Khotan and Yarkand have come pebbles and boulders of the finest material. This was undoubtedly nephrite and was considered true jade for it was not until much later that jadeite found its way from Burma to the Chinese artisans of Canton and Peking. The most important district for Burmese jade is Mogaung in Upper Burma, where it is recovered as cobbles and pebbles from the river Uru.

Jade implements fashioned by primitive peoples have been found in numerous places in Asia, Europe and America. In Mexico and Central America, many

(Above) Rose-red rhodochrosite crystals with galena (black) and pyrite (brass-yellow) implanted on quartz, from a silver mine at Alma, Colorado. (Studio Hartmann)

(Below left) The range in the color of garnets is shown in pink grossularite (left) from Chihuahua, Mexico, brown andradite (center) from Vask, Hungary, and orange-yellow spessartite (right) from Ramona, California.

(Below right) An aggregate of small grains of olivine forms a large nodule broken from a basaltic lava flow. The small crystals and cut stone show the characteristic color of gem olivine, peridot, from Saint John's Island in the Red Sea. (Both by Emil Javorsky)

carved jadeite pieces have been excavated on the sites of Aztec and Mayan civilizations. It played an important part in the religious and social life of such early peoples, and elaborate carvings attest to the skill of the pre-Columbian artisans. It is even claimed that Montezuma considered a gift of two pieces to Cortez equivalent to two cartloads of gold. Jade has not been found in place in Mexico and it is believed that the Aztecs found it as stream pebbles, probably in southern Mexico or Guatemala.

When Captain Cook discovered New Zealand in 1769, he found that the native Maoris, like the Aztecs and Mayas, treasured nephrite jade and carved pendants and earrings from it. They also worked it into knives, axes, adzes and war clubs. On the west coast of the South Island, where the Taramakan and Arahura Rivers empty into the sea, the Maoris obtained the jade as boulders and pebbles in the rivers and along the nine miles of beach between the rivers. Although nephrite has been found in the Griffin Range, that deposit was apparently unknown to the Maoris.

Many minerals, some naturally green and others dyed green, have been used to imitate jade. The most popular is dense massive serpentine. Its jade-green color can mislead the unwary, but it can be detected by its inferior hardness: it can be scratched by a knife whereas jade cannot. More difficult to detect are harder minerals resembling jade such as the massive green idocrase, californite, and the green grossularite, garnet, called "South African jade." Because of the high prices commanded by early carvings, counterfeiting of pre-Columbian pieces using true jade is fairly common. The mineralogist can determine only the substance from which the piece is made, but it requires the expertise of the archeologist to discover such a fraud.

Garnet

Garnet is a name given to an entire series of minerals although the members vary in occurrence, chemical composition and physical properties. The garnet used as a gem for thousands of years is dark-red, so characteristic in color that the term "garnet-red" is used in describing other minerals. Yet garnets are found in all colors except blue, a fact unknown even to many jewelers.

In minerals such as quartz and corundum the chemical composition is essentially constant and color variations result from minute traces of impurities. In garnets a difference in color reflects major differences in chemistry. Although a separate name is given to each chemical variety, one type frequently grades into another. The major types are:

> *Aluminum Garnet*
>   Pyrope: magnesium, aluminum—$Mg_3Al_2(SiO_4)_3$
>   Almandite: iron, aluminum—$Fe_3Al_2(SiO_4)_3$
>   Spessartite: manganese, aluminum—$Mn_3Al_2(SiO_4)_3$
>   Grossularite: calcium, aluminum—$Ca_3Al_2(SiO_4)_3$
>
> *Iron Garnet*
>   Andradite: calcium, iron—$Ca_3Fe_2(SiO_4)_3$
>
> *Chromium Garnet*
>   Uvarovite: calcium, chromium—$Ca_3Cr_2(SiO_4)_3$

Figure 4

Figure 5

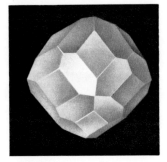

Figure 6
Garnet Crystals

These formulas reveal that in spite of major differences, there are similarities: thus the $(SiO_4)_3$ is common to all. They are all silicates with the same crystal structure but with corresponding places in the structure occupied by different atoms. For example, iron in almandite fills the place of magnesium in pyrope. Moreover, a partial substitution of one element for another, as iron for magnesium, or magnesium for iron, results in a complete gradation from one garnet to another. In no other group of minerals is this phenomenon, known as solid solution, better illustrated than in the garnets.

Since all garnets have the same structure, they all have similar crystal forms. These forms were early thought to resemble the seed of a pomegranate, and it is from *granatus,* meaning pomegranate, that the name garnet is derived. Well-formed garnet crystals are equidimensional, and some, covered with many faces, are nearly spherical. The twelve-sided crystal form, the dodecahedron (Fig. 4), and the twenty-four-sided trapezohedron (Fig. 5), or combinations of these two (Fig. 6), are the most frequent shapes. Well-formed garnet crystals are rather common, for the mineral has an extremely strong crystallizing power and is able to develop its characteristic crystal shapes under adverse conditions.

With the exception of uvarovite, gems have been fashioned from all of the garnet types and are sold under a great variety of names. Some of the names are false and others misleading but most represent color varieties. The garnet most often used in jewelry is the deep-red to nearly black pyrope, the name coming from a Greek word meaning firelike. The ruby-red color of some pyrope has led it to be sold as ruby; thus, "Arizona ruby" is the name sometimes given to pyrope stones picked up in the Arizona desert, and "Cape ruby" to those found with diamonds in South Africa. The best-known source is near Teplitz, Czechoslovakia, from which the so-called Bohemian garnets have come. So abundant were these stones and so popular the jewelry fashioned from them that many assume that their red-brown color is the color of all garnets.

The iron-aluminum garnet, almandite, is a deep-red to a violet-red. These stones cut *en cabochon* and called carbuncle have been known since Biblical times. One of the stones in the "breastplate" of the high priest is believed to have been carbuncle. Almandite of gem quality has come from many parts of the world but the principal source has been the stream gravels of India. Rhodolite is a lovely pale rose-red to pinkish-red garnet, intermediate between pyrope and almandite in chemical composition. Its only notable source is in the United States in Macon County, Georgia.

The name of the mineral grossularite comes from *grossularia,* the botanical name for gooseberry. Although the mineral is found in many colors, it was the pale-green variety, resembling the color of gooseberries, that earned it its name. With other gem minerals from the gravels of Ceylon has come an orange to golden grossularite called *cinnamon stone* or *hessonite*. And north of Johannesburg, South Africa, from the great geologic body known as the Bushveld Igneous Complex, has come the green massive grossularite sold as "South African jade."

Stones cut from the manganese-aluminum garnet spessartite, although of an attractive orange-red color, are not widely known, chiefly because there is so little spessartite of gem quality. Most of the gem spessartite has come from the gravels of Ceylon and from the pegmatites at Amelia Court House, Virginia.

Andradite, the calcium-iron garnet, is a common mineral, but it is usually opaque and thus unsatisfactory for gems. There is an exception to this: one variety is fashioned into superb gems, the most valuable of any of the garnets—green demantoid. When faceted it flashes spectral colors, earning it a name that means diamond-like, although its hardness (6½) is much lower than that of the precious stone. Demantoid was first found in the 1860's near Nizhni-Tagil, Siberia, by gold miners who were panning the stream gravels. It was later located on the western slope of the Ural Mountains in nodules as large as two inches in diameter. Because of its deep-green color, it was first thought to be emerald, and even when its identity was known it was called "Uralian emerald." But this gem need not masquerade as any other stone, for its beauty rivals that of any of them.

Another green garnet, uvarovite, with its deep-emerald color and brilliant luster, would make a superlative gem were it found in crystals large enough to be cut. It is a calcium-chromium garnet occurring only as tiny flashing crystals in thin veins cutting through masses of black chromite.

The occurrence of garnet is as varied as its chemical composition and color. It is abundantly present in metamorphic rocks, schists and gneisses, and as a product of contact metamorphism in crystalline-limestones. It may be a constituent of igneous rocks, and large crystals are found in pegmatites. Because of its resistance to chemical and mechanical weathering, garnet is common in sands, particularly black sands made up of minerals of high specific gravity.

Garnet is an important abrasive material, and over ten thousand tons are produced annually in the United States alone. Only a few garnets are suitable for abrasives since the mineral must not only be hard but it must break under pressure to yield sharp cutting edges. The only major deposit in the United States where garnet has this property is at Gore Mountain in the Adirondacks, where a mine has been producing almandite garnet for abrasives since 1880. The crystals here range in diameter from less than an inch to over a foot, with an occasional crystal as large as three feet. The almandite is surrounded by a rim of black hornblende that sets off the red-brown of the garnet.

Peridot

In antique jewelry one frequently encounters a clear olive-green stone that someone may once have bought as so-called "evening emerald." In all likelihood, the cutter knew it by no other name but today it would be identified as peridot, the gem variety of the mineral olivine. In early times, emerald, in spite of its deeper color and greater hardness, was doubtless often confused with olivine. Peridot makes a most attractive gem and, though relatively soft (hardness 6½) for a ring mount, would no doubt be popular were there a larger supply.

Olivine, sometimes called chrysolite, is a magnesium-iron silicate, one of the rock-making minerals. It is common in basalts as glassy grains scattered through the black rock; it is one of the major constituents in peridotites and the only important mineral in dunites. In all of these rocks, the olivine occurs in small grains frequently cracked and clouded; rarely does one find a clear piece which is large enough to cut a one-carat stone.

Beach sand near South Point in the Kau district of Hawaii is colored green by tiny olivine crystals weathered from lava flows that form the neighboring cliffs. (Werner Stoy: Camera Hawaii)

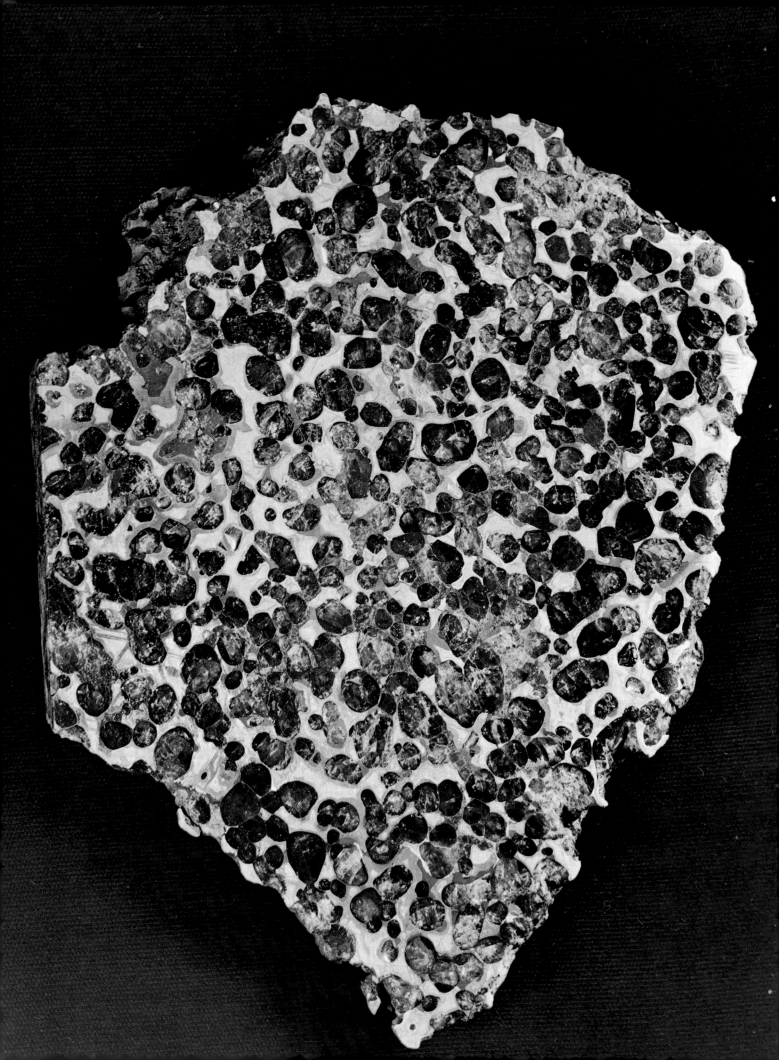

It is thus apparent that the large olivine crystals from which peridots of fifty, one hundred, and even three hundred carats have been cut were formed under unusual conditions. During the nineteenth century the source of the ancient stones was the subject of much speculation for no records of their source existed. Since peridot was known in Biblical times and was used by the ancient Egyptians, it was generally thought that the source was somewhere in the eastern Mediterranean area. In the meteorites known as pallasites a network of metallic iron encloses grains of olivine, some of which are of gem quality. One theory therefore proposed an extraterrestrial origin for the peridot and suggested that the meteorite from which it came had long since been carried away. It was not until early in the twentieth century that the ancient source was rediscovered—on the small island of St. John in the Red Sea, fifty miles from the Egyptian port of Berenice. The rocks of the island are composed essentially of olivine in grains too small for gems, but on the eastern side there are veins containing beautiful, well-formed crystals. The many shallow pits in this area are no doubt the remains of ancient mines. Since their rediscovery, new pits have yielded superb crystals, but the supply is limited.

Other deposits of olivine of gem quality but of lesser importance occur in Burma, Ceylon, Australia and Brazil. In the Navajo Indian country of New Mexico and Arizona, pale-green grains of clear olivine can be found in the surface sand and gravel. They crystallized as a constituent of the underlying rock and have been liberated by weathering; but the small stones that can be cut from them do not compare with the magnificent peridots from St. John's Island.

A polished slice of a pallasite (stony-iron) meteorite. Green olivine grains are enclosed in a matrix of nickel-iron. (Carlton B. Moore, Center for Meteorite Studies, University of Arizona)

17 Minerals for Atomic Energy

The explosion of the atomic bomb at Hiroshima in 1945 ushered in the age of atomic energy and with it an intensive search for uranium as the fuel for that energy. There are approximately one hundred minerals in which uranium is an essential constituent, but only five or six are abundant enough to be major ores of the element. More than half of these minerals, unknown a generation ago, were found and described as a result of the unprecedented world-wide search for uranium following World War II.

Although minerals containing uranium have been known for 150 years, they were considered worthless, and for a century were mined only because of associated metals, chiefly silver, cobalt, gold and copper. Following the discovery of radium in 1898, uranium minerals became important as the ores of this element. But uranium contains very little radium: thus, approximately 21,000 tons of ore must be mined to get 336 tons of concentrate, and after chemical treatment this yields only one ounce of a radium salt! Most of the uranium by-product of the radium industry was waste but small amounts were used in the textile and ceramic industries, as an alloy in steel, and in other minor ways. With the great demand for uranium as the fuel for atomic energy, these early uses are now nearly forgotten.

How does uranium differ from other elements and why should it have such potential power? The answers lie in the structure of its atoms. Uranium and a few other elements are endowed with a property, radioactivity, that results from the spontaneous breakdown of their atoms. The atoms of the chemical elements are all made up of a nucleus around which electrons revolve much as the planets revolve around the sun. The weight of an atom is concentrated in the small nucleus, which is composed essentially of two types of particles—positively charged protons, and neutrons with no charge. Each of the electrons

(Above) Orange-yellow gummite from Grafton Center, New Hampshire, has formed as an alteration product of black uraninite.
(Benjamin M. Shaub)

(Below) Tabular crystals of green autunite from Spokane, Washington, have grown together into fanlike aggregates.
(Floyd R. Getsinger)

spinning around the nucleus carries a negative charge and since the number of electrons equals the number of protons, the atom as a whole is electrically neutral.

Each element has an atomic number that is equal to the number both of its protons and of its electrons. Hydrogen, the simplest and lightest of the elements, has the atomic number 1, and each of its atoms is composed of a nucleus containing one proton around which is spinning a single electron. With increasing atomic number the atoms of the elements became heavier and more complex. Uranium, the heaviest of the elements found in nature (atomic number 92), is composed of atoms whose nuclei contain 92 protons and 146 neutrons. With so many neutrons and protons the nucleus breaks down and endeavors to adjust itself to a more stable condition. It does this by emitting an alpha particle, the nucleus of a helium atom with $+2$ charge. This gives rise to the nucleus of the thorium atom, whose nucleus has a charge of $+90$. The thorium atom is also radioactive and in turn emits a beta particle, forming the nucleus of another element. Radioactive disintegration results in a whole family of elements, starting with uranium; one offspring gives rise to another until the stable atom of lead (atomic number 82) is reached.

During this process of "radioactive decay," three types of radiation are emitted: alpha particles, beta particles and gamma rays. The alpha particles are nuclei of helium atoms and have two protons and two neutrons. They are emitted at such great velocities that they will travel through as much as three inches of air before their energy is spent in knocking off electrons from the atoms of oxygen and nitrogen with which they came in contact. Beta particles are electrons traveling with a velocity almost as great as that of light. They too knock off electrons from the atoms of gases through which they pass but they are somewhat more penetrating than alpha particles. Gamma rays, are similar to x-rays and like them travel with the speed of light, but they differ from most x-rays in having a shorter wave length and greater penetrating power. Gamma rays are chiefly responsible for the darkening of photographic film, and they can pass through a half-inch of lead without losing more than half their intensity. Radioactive disintegration also generates heat.

Although all atoms of an element have the same number of protons and electrons, they may differ in the number of neutrons and thus have different weights. Most uranium has an atomic weight of 238 but about one atom in 140 has an atomic weight of 235. It is this small amount of U^{235} which, when separated from the U^{238}, is the major source of atomic energy. When the heavy nucleus of U^{235} is bombarded by neutrons, the atom splits. This fission forms two atoms of lighter elements, producing from one to three neutrons and releasing energy. The neutrons will cause splitting of other nuclei, setting up a chain reaction that altogether liberates a tremendous amount of energy. Since only the nucleus is involved, atomic energy is also called nuclear energy.

The rapidity with which different radioactive elements disintegrate varies greatly; the greater the radioactivity, the faster the decay. But the rate for any element is constant, regardless of temperature and pressure. It is given in terms of its "half life," that is, the time required for half of the original atoms to disintegrate; a like amount of time will then be necessary to disintegrate half of the remaining atoms. The half life of uranium-238 is 4,500,000,000 years, of thorium-234 it is 24.1 days, and of polonium-214 less than one-thousandth

The brilliant colors of the secondary uranium minerals are illustrated by specimens from Katanga: (above left) yellow kasolite, reddish-orange cuprite; (above right) yellow dewindite, brownish-yellow kasolite; (below left) green vandenbrandeite and yellow soddyite; (below right) bright-yellow carnotite with chervetite, a lead vanadate. (Emil Javorsky)

Mining operations by hand in the sixteenth century are depicted in Mining in the Alps *(ca. 1530) by Hans Holbein. (The British Museum, London)*

of a second. By knowing the present quantity of the parent element and its rate of decay and the quantity of the end product, geologists can calculate the age of a radioactive mineral and of the igneous rocks in which it is found.

Detecting Radioactivity

The radioactivity of uranium minerals enables the mineralogist to detect them in the laboratory and the prospector to find them in the field. The radiation emitted during radioactive decay, particularly gamma rays, will darken a photographic film; thus, if a specimen containing a radioactive mineral is placed on a photographic film inside a light-tight envelope, it will take its own picture—an autoradiograph. By comparing an autoradiograph with the specimen, one can quickly tell which mineral grains are radioactive.

The best-known device for detecting radioactivity is a Geiger-Müller counter, known as the Geiger counter. It is a glass tube fitted with a cylindrical cathode and a wire anode at the axis of the cylinder and filled with a dry gas at about one-tenth of atmospheric pressure. When products of radioactive disintegration enter the tube they ionize the gas, causing it to produce a momentary electric discharge. By amplification the discharge produces a "click" in a loudspeaker, a flash of light, or a "count" in a counting device.

The greater the concentration of radioactive material, the more rapid the succession of clicks and flashes and the more counts in a given time. Portable Geiger counters are available and with such instruments even the untrained person can detect the presence of uranium minerals.

The scintillation counter, another detecting device, is more expensive than a Geiger counter, but it is more sensitive and will record gamma rays at great distances. With it a radioactive source can be detected even from an airplane; if this instrument is flown over large uncharted areas it can help determine favorable places for closer examination.

Primary Uranium Minerals

Of the many uranium minerals only a few are primary, and of these only uraninite, uranium oxide, is an important ore. Moreover, as a result of its alteration, the great host of secondary uranium minerals is formed. Under ideal conditions for crystallization, uraninite forms in black, cubic or octahedral crystals rarely more than two inches across. It may be found in scattered grains in granites, but well-formed crystals are very rare and come mostly from pegmatites. Uraninite is most commonly a vein mineral in massive or botryoidal aggregates with a pitchlike luster called pitchblende, and is associated with ores of silver, cobalt, nickel and copper.

Uranium was discovered in 1789 by the German chemist, Martin Klaproth, in pitchblende from the silver mines at Johanngeorgenstadt in the Erzgebirge of Saxony. He called the element uranium and the mineral from which it came, uraninite. It was also from pitchblende mined across the mountains at Joachimsthal, Bohemia (Czechoslovakia), that Pierre and Marie Curie in 1898 isolated the element radium. This area not only yielded ores for the early production of radium but is known to have furnished large quantities of uraninite for the nuclear weapons of the Soviets.

Secondary Uranium Minerals

In contrast to the somber blacks of the primary uranium minerals, the secondary uranium minerals are a variety of brilliant shades of yellow, orange and green. They frequently occur in small delicate crystal plates or needles or in earthy or powdery masses. No other element, with the possible exception of copper, forms a more beautiful and attractive group of minerals. Many of them also fluoresce under ultraviolet light, increasing their appeal to the mineral collector.

These minerals are called secondary because they are formed at the expense of the primary minerals. They may occur near the surface in oxidized zones overlying the primary deposits from which they have formed, or they may be found in flat-lying sedimentary rocks deposited from solutions that have carried uranium far from its primary source. The brilliant colors of secondary uranium minerals help a prospector locate deposits, but they may also cause his hopes to rise too sharply, for some are strong pigmenting agents and it requires only a small amount of bright-yellow carnotite, for example, to color a large volume of rock.

The most important secondary uranium minerals are listed in the following table.

Secondary Uranium Minerals

	Color	Crystal	Fluorescent Color
Carnotite	lemon-yellow	powder	nonfluorescent
Tyuyamunite	greenish-yellow	powder	nonfluorescent or yellow-green (weak)
Torbernite	emerald-green	flat, square tablets	faint-green
Autunite	sulfur-yellow	thin, square tablets	yellow-green
Uranophane	pale-yellow to orange-yellow	small needles	pale-green or nonfluorescent
Schroeckingerite	greenish-yellow	flattened scales	yellowish-green

Primary Deposits

Although the localities in which secondary uranium minerals are found far outnumber those of primary minerals, it is the primary deposits of uraninite that have been most productive. The three most famous that have supplied most of the world's uranium are the Erzgebirge district of Germany and Czechoslovakia; Shinkolobwe, in the Congo, and Port Radium in the Northwest Territories of Canada. Mining in the Erzgebirge has continued almost without interruption since the sixteenth century. At first the area was mined for native silver; it was not until the nineteenth century that the associated pitchblende was sought out as a ceramic coloring agent. The most important uranium mines in the district are at Joachimsthal, Czechoslovakia, where uraninite is found in veins along with minerals of silver, cobalt, nickel and bismuth. The commonest surface sign of uranium here is rock outcrops stained yellow and green by autunite and torbernite.

The Shinkolobwe mine lies in Katanga in the southern Congo not far from the rich mineral deposits of the "Copper Belt" of Zambia. Primitive copper utensils indicate that the natives knew of copper long ago. It was in the search for copper that prospectors in 1915 came upon brilliantly colored rock fragments which proved on analysis to be rich in uranium. As the prospecting trenches went deeper, the fragments became increasingly abundant until, at bedrock, veins containing uraninite were exposed. Mining of uranium as an ore of radium began in 1921; since that time Shinkolobwe has produced more uranium than all other world deposits combined. As the work at Shinkolobwe progressed, it was found, moreover, that the ore body was composed of a network of pitchblende veins that also contained minerals of cobalt, copper, tungsten, gold and platinum. In the subtropical climate of Katanga deep weathering of the primary pitchblende has produced secondary minerals to a depth of several hundred feet. Near the surface, abundant torbernite, autunite, uranophane and a score of other brightly colored secondary minerals have been found. Some veins show only these alterations products at the surface but in others pitchblende makes up the central portion bounded by secondary minerals in tabular zones.

Since much of the Canadian Northwest Territories is a wilderness, with transportation largely by canoe, it is not surprising that prospectors early took to the air to scan large tracts for promising signs of minerals. It was on such a reconnaissance trip that Gilbert La Bine discovered the Eldorado deposit in

1930. Flying over the eastern end of Great Bear Lake he noted bright-colored outcrops on an island and, a short distance away, similar evidence on the shore of the mainland. On the island he discovered silver and cobalt minerals and on the mainland a vein not only of these minerals but of pitchblende as well. Further prospecting revealed that this was one of five highly mineralized veins, each several thousand feet long and in places as much as thirty feet wide. The advancing ice of the last glaciation had removed any great thickness of secondary uranium minerals that may have once been present. Nevertheless, surface alteration of the pitchblende since glacial times had formed the bright-yellow, green and orange secondary uranium minerals on the outcrops of the veins. Mining was begun at Port Radium, first for native silver and later for uranium as a source of radium, but it was not until the early 1940's that uraninite was taken out as a source of uranium itself.

The physiographic unit of the United States known as the Colorado Plateau embraces an area, mostly in the Southwest, of about 150,000 square miles. Probably the first uranium mineral to attract attention in this region was carnotite, a bright yellow mineral—an ore of vanadium as well as uranium—used by American Indians as a war and ceremonial paint. Commercial production of carnotite on the Colorado Plateau began in 1910 and the mineral was shipped to France as an ore of radium until 1924 when it became impossible to compete with the high-grade ores from the Congo.

Much of the great search for uranium in the United States following World War II took place on the Colorado Plateau. Hundreds of prospectors walked thousands of miles examining the sandstone outcrops in valley bottoms and high mesas. Carnotite was the mineral they most often located. Deposits may be only a local, powdery crust or a concentration several feet wide and several hundred feet long; they may be tabular and of uniform thickness, following the bedding of the enclosing sandstone, or lens-shaped masses wide at the center and tapering to a thin seam at the edge. The enclosing rock characteristically contains abundant plant remains and fossil logs. This organic matter appears to have influenced the precipitation of uranium minerals from solution, and carnotite frequently is the petrifying material. From one giant log, one hundred feet long and four feet in diameter, miners recovered over one hundred tons of ore valued at $230,000.

Prospecting reached a peak in 1954–55 when two thousand individual deposits, most of them small, were being worked on the Colorado Plateau. Although in the following years the number of producers decreased, the ore mined steadily increased, reaching eight million tons in 1961. From that time production has fallen but, with the ever growing demand for uranium for peaceful atomic energy uses, mining of the major deposits will continue for many years.

18 The Mines and Minerals of Cornwall

Probably no other metal-mining area in the world has had as long a productive life as Cornwall, the small county at the southwestern tip of England. Copper, lead, zinc, wolfram, iron and uranium have been produced here in significant amounts; but it was tin that first attracted miners to its streams and merchants to its shores and has made it a world-famous mining district. Mining is believed to have begun between 1800 and 1500 B.C. The Romans were active miners, but probably more than one thousand years before the Roman period in Britain the Phoenicians sailed to Cornwall to trade pottery, salt, and metal implements for tin.

The first use of tin by man was in the ancient bronzes of the "Bronze Age." Archeologists cannot give an exact date when the Stone Age passed into the Bronze Age for it differed from place to place. Bronze articles believed to be the earliest were found in Egypt and date from the IV Dynasty (4700–3800 B.C., depending on the authority). Other less advanced cultures may not have known of bronze until more than two thousand years later.

Bronze is an alloy of copper and tin; although the ratio between the two metals varies greatly, copper is always dominant. It seems probable that bronze was discovered accidentally when someone smelted a mixture of ores from a locality where copper and tin minerals occurred together. The source of these ores is unknown and there is, in fact, disagreement even as to whether they were found in Africa, Asia or Europe. However, there is general agreement that by early in the second millenium B.C. bronze was being used by all the peoples on the shores of the eastern Mediterranean. It is also thought that by this time the ancients knew that bronze was an alloy of copper and tin and could be made by melting the two metals together. In comparison with tin, copper is abundant, and many localities in the area bordering on the Eastern

Smokestacks and stone shells of ancient pump houses are all that remain of the once thriving tin mining industry near Land's End, Cornwall, England.
(Werner Lieber)

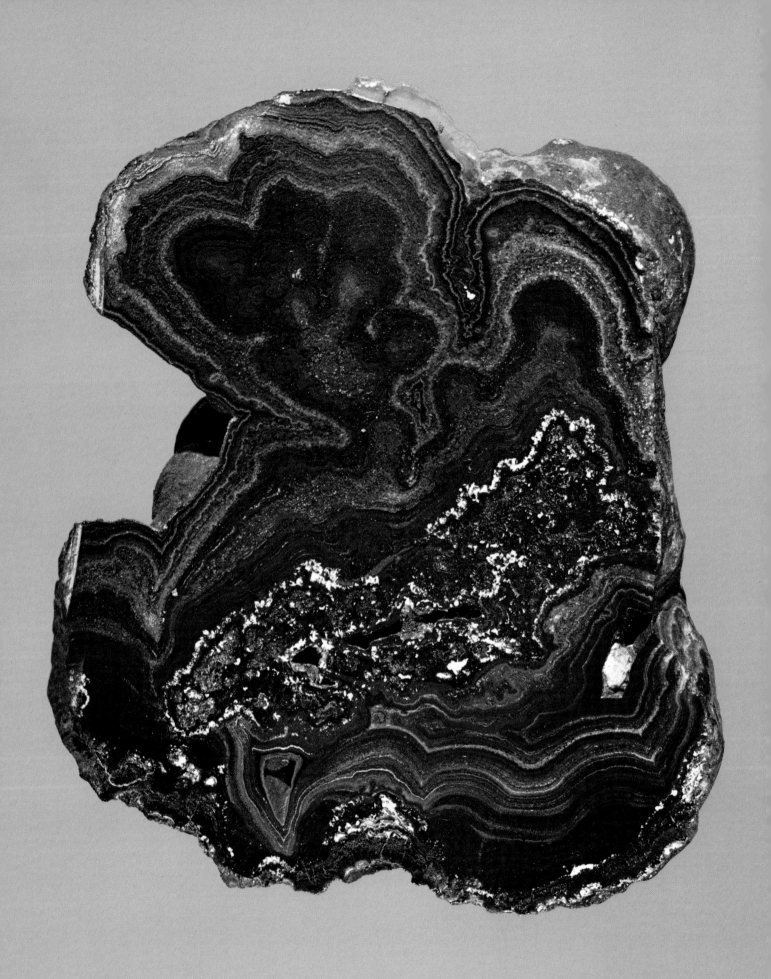

Mediterranean could have furnished this metal. However, there is little evidence that tin was available in the region and its source has perplexed historians.

Ancient Phoenicia lay at the eastern end of the Mediterranean along a strip of coastal land now occupied by parts of Syria, Lebanon and Israel. With ample timber in the hills to the east the Phoenicians built ships and became the foremost mariners of their time. Unlike many of their contemporaries they were interested not in conquest but in commerce, and between 1200 and 700 B.C. they dominated international trade. About 1100 B.C. they founded the city of Gades on the site of present Cadiz. Gades became a tin market to which tin was brought from islands in the Atlantic known as the Cassiterides, a name taken from the Greek word meaning tin. The location of these islands has long been debated, partly because of the Phoenician success in keeping secret the source of the tin and thus retaining this profitable trade for themselves. Some think that the Cassiterides were off the Spanish coast, but it now seems almost certain that the Phoenicians sailed their primitive crafts one thousand miles north from Gades and obtained tin from the natives on the southern coast of Cornwall. It was from the same shores that the Romans later obtained tin. Their record is more complete and we learn that the tin was transported across the English Channel and then taken overland to Marseilles.

The tin mineral mined by both early man and his more sophisticated descendants was cassiterite, tin oxide (SnO_2). Cassiterite occurs in Cornwall in veins or lodes tenaciously held in place by the enclosing rock. Primitive miners working only with tools of wood and deerhorn could not possibly have extracted it. However, nature working gently over long periods of time disintegrated the surrounding rock, setting the cassiterite free. Cassiterite is chemically inert and not susceptible to chemical attack by weathering agents. Furthermore, it is hard ($H = 7$) and has a high specific gravity (7.0), and thus has all the attributes of a mineral to be collected in placers.

During the glacial period the flood waters washed disintegrated rock into the streams, carrying to sea some of the lighter minerals but leaving a concentration of heavier ore minerals in the lower reaches of the streams. It was in the gravels of these ancient stream channels that the early miner looked for pebbles of "stream-tin" that nature had concentrated for him. Veins containing cassiterite were probably located in early times but the first hard-rock mining in Cornwall is believed to have been by the Romans at what became known as the Ding Dong mine on the Land's End peninsula.

Mineral Deposition at Cornwall

The minerals of Cornwall that have attracted the attention of the miner and mineralogist are associated with coarse-grained granite intruded into the older sedimentary rocks toward the close of the Carboniferous period. The granite is exposed in five major masses, each five to ten miles across, that trend northeast from Land's End to Exeter in Devon.

During the crystallization of the intruded magma, the elements required by the rock-forming minerals, feldspar, quartz and mica, were slowly withdrawn from the molten mass, while other elements unnecessary for their development became concentrated in the noncrystalline fluid portion. Water comprised a high percentage of this material but fluorine, boron, sulfur dioxide and carbon

Cassiterite from San Luis Potosi, Mexico, shows the concentric banding characteristic of a variety known as wood tin. *(Benjamin M. Shaub)*

dioxide were present as well. Also to a lesser extent were the metals tin, tungsten, copper, uranium, iron, zinc and lead. After much of the upper part of the granite had crystallized, gases enriched in these volatile components and rarer elements emanated from the magma. They rose through fractures in the overlying granite and neighboring rocks, giving rise to metallic lodes and altering the rocks with which they came in contact.

In the earliest stage of alteration boron was a major ingredient in the solutions, and the granite became "tourmalinized." Although some tourmaline is clear, pleasingly colored and cut as gem stones, the kind formed by this process is black and appears opaque. In this process the mica was first attacked by the hot rising fluids and its place taken by tourmaline; but as the process continued, the feldspar was also replaced and a quartz-tourmaline rock was formed. In addition to the alteration of the rock-forming minerals, in some places veins of quartz-tourmaline traverse the granite in all directions.

In the second stage of alteration (called "greisening" by British geologists), fluorine is the dominant active ingredient in the solutions and the rock is replaced by minerals containing that element. Fluorite, CaF_2 and topaz, $Al_2(SiO_4)(F,OH)_2$ were deposited, sometimes in well-formed crystals. At the same time much of the feldspar of the granite was converted into quartz and into a white mica that may contain lithium and fluorine. Locally one can find crystals, originally feldspar, completely altered to topaz but maintaining the crystal forms of the feldspar. They are pseudomorphs of topaz after feldspar.

Kaolinization, the conversion of feldspar to chalky white kaolinite through the action of carbonic acid, was the third and last stage of alteration of the granite. The kaolinized feldspar grains are so soft and so friable that they can be easily broken with the fingers. In places the large feldspar crystals may be only partly kaolinized but elsewhere they have been completely converted into pseudomorphs of kaolinite after feldspar.

The process of kaolinization is in most places in the world a surface alteration. The rain water, containing dissolved carbon dioxide, attacks the feldspars, slowly converting them to kaolinite and other clay minerals. The soils over much of the earth's surface have been formed in this way. Such alteration is usually shallow and gives way to fresh rock at a few feet to a few tens of feet beneath the surface. At Cornwall surface alteration is negligible compared with the deeper kaolinization resulting from hot vapors ascending through the rock. This type of alteration is shown particularly well in the St. Austell granite where kaolinization is not uniform but occurs in definite elongated northeast-southwest zones, indicating that it resulted from vapors rising through fissures trending in this direction.

Although kaolinization has effected all of the granite of Cornwall, it is most extensive in the granite mass of St. Austell. Here nearly pure kaolinite has been recovered as china clay since the middle of the eighteenth century. In a simple but effective process, called hydraulicing, clay is separated from the worthless minerals of the granite. In this method jets of water under high pressure are directed against the walls of a pit, breaking up and washing the soft, kaolinized rock to the bottom. The large particles of quartz and other granular minerals quickly settle out of the water. But the extremely small flakes of kaolinite as well as the somewhat larger flakes of muscovite mica are held in suspension, and are pumped to settling tanks at the surface. In the first tank the granular waste material settles and the water carrying the flocculent mica and kaolinite is drawn off into a second tank. Here, because of its larger flakes and higher specific gravity, the mica settles, leaving the kaolinite still in suspension in the water. Eventually, the kaolinite settles and is dug out and placed on drying floors. When dry, it is ready for manufacture into high-grade chinaware. The waste material from which the mica and clay have been removed is composed largely of quartz but also contains the other minerals of the granite such as unaltered feldspar, tourmaline and topaz. In the area of St. Austell high conical waste piles of glistening quartz, accumulated through two centuries, form landmarks that can be seen for many miles.

When the early miners traced stream tin to outcrops in the hills, they found cassiterite in ore bodies along with minerals of several other metals. Studies of these lodes have shown that the ore minerals were deposited from solutions that emanated from the crystallizing magma and passed through fissures in and around the granite. Associated with the ore minerals and deposited from the same solutions are such minerals as quartz, barite, fluorite, tourmaline and chlorite. These are the gangue, or gangue minerals, and contain no recoverable metal.

The ore bodies of Cornwall are arranged in a crude but definite series of zones circling the granite masses. As the solutions moved away from the hot crystallizing granite, they became cooler and with the falling temperature deposited a succession of minerals. Four major zones have been distinguished with one or two metals dominant in each. But since the process of vein formation was continuous, there is overlap between the several zones.

The first zone, closest to the granite and thus formed at the highest temperature, is characterized by abundant cassiterite. Tungsten minerals, particularly wolframite, are also present and grade into the next zone, that of the copper sulfides. Accompanying the copper sulfides, chalcopyrite and bornite, is pyrite, uraninite, and minerals of cobalt, nickel and arsenic. The copper zone grades into a zinc-lead zone. First zinc sulfide, sphalerite, makes its appearance and as the copper sulfides diminish, the lead sulfide, galena, and silver minerals are found. Finally, farthest from the granite, is an iron-rich zone with the iron carbonate, siderite, a dominant mineral. In addition to this type of successive mineral deposition proceeding outward from the granite, individual mines show a similar vertical zoning where the mineralogy changes with depth. For example, the Dolocath mine was in its early days a producer of zinc and copper but ended as a tin mine. The vertical distribution of the metals was as follows: Zinc and copper occurred together in the upper levels, but copper alone continued to one thousand feet. Between one thousand and 1150 feet both copper and tin were found and from 1150 feet to the bottom of the mine, 3300 feet down, tin alone was mined.

Accompanying the zoning in the metals there is a zoning in the associated grangue minerals. Quartz has no favorites: it is found everywhere from the innermost to outermost zone and from the surface to the bottom of the deepest mine, but other minerals differ from zone to zone. Tourmaline is a common and characteristic associate of cassiterite in the tin zone but as copper minerals become more abundant it gives way to fluorite. Barite is found with the lead and zinc ores, and calcite and siderite are associated in the zones farthest from the granite and in the upper parts of some lodes.

New Minerals from Old

When ore minerals are exposed at the surface to the action of rain water with dissolved oxygen and carbon dioxide, various chemical reactions take place. The original minerals dissolve and disappear but the resulting solutions may react with the minerals they encounter as they pass downward to form new compounds. The secondary minerals thus formed contain the metals of the primary minerals that were subjected to surface alteration. If iron is present in an ore body, the commonest surface expression is a gossan composed of the yellow-brown iron oxide, goethite, mixed with gangue minerals. In Cornwall the gossan may extend to a depth of nine hundred feet where it gradually passes into a zone in which secondary copper minerals are present. These are the red copper oxide, cuprite (Cu_2O), the black copper oxide, tenorite (CuO), and malachite and azurite, the green and blue copper carbonates. These are colorful minerals, and fine specimens of them are found in many mineral collections.

In some of the Cornish mines, particularly at Trenwith and South Terras, brilliantly colored secondary uranium minerals have been found associated with the oxidized copper ores. Although we associate uranium and its ores with the atomic age, the uranium minerals of Cornwall were known a century earlier. Historically they are of interest since they and similar minerals from Saxony and Bohemia were the only known sources of uranium until the discovery of the carnotite deposits on the Colorado Plateau in 1898.

As the dilute solutions move slowly downward, much of their metallic content may be precipitated in the zone of oxidation, but some may remain when the solutions reach or penetrate the water table. Oxidizing conditions no longer exist here and metals carried in solution are precipitated as metallic sulfides. The metal carried from above is thus added to the metal already there forming a narrow but rich zone of secondary enrichment.

In copper deposits the mineral chalcocite (Cu_2S) is the principal secondary ore mineral found usually replacing the primary copper minerals, chalcopyrite and bornite. The bottom of the zone of secondary enrichment is less well defined than the top, and grades irregularly downward into the unaltered primary ore.

The Rise and Fall of Cornish Tin

The record of mining in Cornwall from the end of the Roman period, about A.D. 300, until the Norman Conquest is obscure and fragmentary. Nevertheless, there is some evidence that tin was mined at least intermittently during this time. It was not until the middle of the twelfth century A.D. that the documentary history of Cornwall and the records of tin production are available. Although little was written regarding the way tin was recovered, it was probably won mostly from alluvial deposits as it had been two thousand years earlier. The workers were called "tinners" in contrast to the "miners" who later worked the lode deposits. As early as the year 1194 the "stannaries" (tin mines) attracted the attention of the crown and for several centuries stannary taxes were a major source of royal revenue.

Although there is evidence that some underground mining was carried out in ancient times, lode mining did not begin in a major way until the beginning of the seventeenth century. There was not an abrupt change but rather a gradual transition over two centuries from "streaming" to hard-rock mining as the easily recovered cassiterite of the streams became depleted. Nor were the lodes first exploited by means of shafts but rather in surface trenches and on the faces of cliffs where the valuable minerals were easily seen and exposed to the light of day. The problems of mining increased many fold when it became necessary to reach the ore bodies by means of shafts. With only flickering candles for illumination, the miner worked in semi-darkness in cramped quarters and poor air, with his clothes wet if not completely soaked by water dripping from overhead. The water was not only a discomfort to the miner, but limited the depth to which the shaft could be sunk. At first the groundwater seeping into the mine workings and accumulating in the bottom of the mine was raised to the surface by manual labor, later by a "horse whym" in which horses were used to raise water-filled barrels, and still later by water wheels, the motive power of which is unknown. As the mines deepened, none of these methods could remove the water as fast as it accumulated, and mining ceased.

It was the steam engine that came to the rescue of the Cornish mines. Although the exact data of the introduction of the steam-pump is uncertain, the Newcomen engine came into use early in the eighteenth century. But, it was an inefficient machine and was replaced in 1778 by the more economical and powerful Watt engine. James Watt's installation of his first pumping engine at the Ting Tang mine at the base of Carn Marth signaled a new era in Cornish mining. With only slight modifications this type of engine is still used today. With it, water can be raised from the deepest mine, the depth of which is determined by the available ore not the flow of water.

With the introduction of the steam pump, every mine had its engine-house, a tall slender building built of stone surmounted by a smoke stack. In the past two hundred years many a Cornish mine has been discovered, flourished and disappeared, leaving only the engine-house as a solitary landmark to indicate its former presence.

During its long productive history, hundreds of mines have operated in Cornwall, thousands of miles of tunnels have been driven and each decade has yielded more metal than the previous one. The first half of the nineteenth century was the most productive period, with the peak of prosperity reached in the 1850's. Cornwall led the world in the production of copper and tin with more than half of its revenue coming from copper. But with the discovery of copper in other parts of the world, particularly in the Lake Superior district of the United States, the price of the metal fell. Unable to meet this new competition, many marginal mines in Cornwall were closed. Although mining continued there, its preeminent place as a copper producer was ended and it never rebounded to its early nineteenth century peak. As a major world source of tin, its ore reserve was much depleted during World War I and during the following years production was gradually decreased. In 1927, A. K. H. Jenkin, in *The Cornish Miner* compared the production figures of 1837 and 1914. In 1837, seven thousand tons of tin were produced from two hundred small mines, whereas in 1914, 6500 tons of tin were produced from ten large mines. From this he concludes: "These figures show clearly enough how comparatively small, according to modern standards, has been the production of Cornish tin in the past and confirm the statements of many eminent mining men and geologists, who claim that, in spite of centuries of working, more tin remains in Cornwall today than has ever been taken out of it."

Notwithstanding this optimistic statement, it appears that the days of Cornwall as a metal mining district are numbered. One by one, mines have been abandoned until in 1960 only one tin mine, South Crofte, remained a major producer. The closing of mines is not only reflected in the economy of Cornwall but in the landscape as well. With the end of mining at many localities, the machinery and the buildings that housed it have been removed. The old landmarks, the pump-house and head frame, are gone. All that remains to mark the presence of a once prosperous mining community humming with both human and mechanical activity are piles of red rock and tailings from which the metals have long since been extracted.

Although time has destroyed the surface evidence, stories of abandoned mines linger on in the memory of the Cornishman. He may remember the Botallack mine where his grandfather worked as a boy of sixteen. Here the lode, originally located on the sea cliffs of the parish of St. Just, followed fissures which plunged beneath the ocean to the northwest. Before the mine was abandoned in the late nineteenth century, the workings extended over three-quarters of a mile out to sea.

Another remarkable mining venture in Cornwall was the development of the Wherry mine in Penzance. Early in the eighteenth century, cassiterite was discovered on a shoal of rocks that was seven hundred feet from shore and was exposed only briefly at low tide. Efforts to mine it were unsuccessful until 1778 when Thomas Curtis renewed the attempt with a bold plan. For three years, working only at low tide, he labored to build a wooden framework cemented with pitch and oakum, the top of which extended well above high water. Behind this bulwark, he sank a shaft which eventually extended to a depth of over one hundred feet. In spite of the hazards involved and operations limited only to the summer, the Wherry proved a profitable mine. But at the peak of its prosperity in 1798 and with high-grade ore still within easy reach, it came to an untimely end when an American ship, blown from its moorings during a storm, struck and demolished the protective framework.

The Cornish miner is as great a contribution to the mining industry as the wealth of metals he has produced. When in the latter half of the nineteenth century he was unable to find employment at home, the miner found opportunities on other continents, carrying with him an invaluable knowledge of practical mining methods handed down through countless generations. A. K. H. Jenkin writes in *The Cornish Miner:* "Wherever a hole is sunk—no matter in what corner of the globe—you will be sure to find a Cornishman at the bottom of it, searching for metal. From Nova Zembla to New Zealand, from Cape Horn to Korea, from Klondike to Cape Town; frozen in the Arctic snows, dried to the bone in tropical deserts, burnt out with fever in equatorial swamps, and broiled thin under equatorial suns—in every country of the New World and the Old the Cornish miner may be found at work." He is still there, wresting metals from the rocks and sharing with the miners of his adopted land the heritage he acquired in the mines of Cornwall.

Near St. Austell, Cornwall, England, clay is recovered from pits sunk in the altered granite, and great piles of waste material, mostly quartz, can be seen for miles. (Cornelius S. Hurlbut, Jr.)

19 An Incombustible Fabric and a Stone That Burns

The most characteristic objects of modern civilization, such as skyscrapers, super highways, automobiles, jet airplanes and atomic energy are made possible only through the use of minerals. We cannot say that one mineral is more important than another for the totality results from the interplay of scores of minerals. Some of the preceding chapters dealt with ores of the metals. In other chapters we discussed the occurrences, associations and changing uses of nonmetallic minerals that also play a key role in modern industry. Two of the most important of these are asbestos and sulfur, each of which is the basic raw material for a great industry.

Asbestos: The "Immortal Linen"

At the height of his power, the emperor Charlemagne, after banqueting his friends, delighted to mystify them by throwing the tablecloth into the fire where it was cleansed of debris without the cloth itself being consumed. The secret of this piece of magic was simple: the cloth was made of asbestos.

The fact that incombustible mineral fibers could be spun into thread and woven into cloth was known long before the time of Charlemagne. *Linum vivum*, "immortal linen," was used by the Romans to wrap the bodies of distinguished persons before cremation in order to preserve their ashes. Asbestos is reported to have been woven into mats by the early Egyptians and Chinese. Outside of this, asbestos appears to have been little used throughout most of historic time. It remained essentially a curiosity until the end of the nineteenth century.

Asbestos is not the name of a single mineral but is the term applied to the fibrous varieties of several minerals of widely differing chemical composition.

A large fragment of crocidolite, "blue asbestos," from Kurman, South Africa. In places the asbestos is completely replaced by quartz but the silky sheen is preserved, as shown in the three smaller pieces of "tiger's-eye".
(Emil Javorsky)

The most important is chrysotile, a fibrous variety of serpentine; the others, all members of the amphibole group, are anthophyllite, crocidolite, tremolite, actinolite, and amosite. Such fibrous minerals are all products of metamorphism but formed in different kinds of rocks and under different conditions.

Tremolite is white calcium-magnesium silicate, in which iron may take the place of some of the magnesium, giving it a pale-green color. With increasing amounts of iron the mineral becomes dark-green and is called actinolite. Thus the asbestos varieties may be white or various shades of green. It is generally believed that these minerals were the first to be given the name *asbestos* and they have therefore sometimes been called "true asbestos." Asbestos usually occurs in "cross fiber" veins from which the individual silky fibers can be easily separated. Some fibrous tremolite, although of no commercial value, occurs in the interesting intergrowths called "mountain leather" or "mountain cork." These mineralogical curiosities are composed of a felted aggregate of interlocking fibers resembling leather or cork. The "leather" type occurs in thin, tough, flexible sheets and the "cork" in thicker but light masses with a spongy, corklike quality.

Anthophyllite is an iron-magnesium amphibole that occurs in rather long, coarse and usually brittle fibers. In 1907, an iron-rich anthophyllite was discovered in the Transvaal, South Africa. The Asbestos Mines of South Africa was the company most interested in its exploitation, and amosite, the name given the fiber, was derived from the initials of the company name.

Crocidolite, sometimes called "blue asbestos" from the color of the fiber, occurs chiefly in South Africa although it is also mined in Australia. It is an iron-sodium amphibole found in cross fiber veins in a rusty colored metamorphosed sedimentary rock. In some places in the Cape Province crocidolite has been replaced by quartz, which retains the fibrous nature of the crocidolite, creating a pleasing effect in reflected light. Cut and polished this material is used extensively in inexpensive jewelry under the name of "tiger's-eye."

Although interesting mineralogically, the amphibole varieties of asbestos are of only minor importance compared to chrysotile, which comprises over ninety per cent of the world's production. Chrysotile is the fibrous variety of the magnesium silicate known as serpentine, and is usually found in the metamorphic rock also called serpentine. Chrysotile usually occurs in cross fiber veins with the fibers ranging in length from one-eighth of an inch to two inches. But in the same rock a "slip fiber" may also be present in which the fibers are parallel rather than perpendicular to the veins in which they occur.

Although asbestos was discovered in 1860 in the serpentine rocks of Quebec, Canada, it was not until ten years later with the finding of the rich and now famous deposits near Thetford Mines and Caleraine that mining began. Although deposits have been discovered in other countries, Quebec, producing over one million tons a year, remains the leading producer of asbestos. However, large deposits of chrysotile, very similar to the Canadian, are being mined on the eastern slopes of the Ural Mountains in the Bajenova district, and with the development of other deposits in Siberia the Russian production may soon equal that of Canada. The Republic of South Africa, Rhodesia and China also produce chrysotile asbestos.

Although the asbestos industry began in the nineteenth century, its major developments have taken place in the twentieth century. Today a host of

(Above) Horizontal veins of gray chrysotile asbestos are exposed on a vertical cliff at Dale's Gorge, Hamersley Range, N.W. Australia.

(Below left) The silky appearance of the cross fiber veins of chrysotile asbestos can be seen in a close-up of the cliff at Dale's Gorge. (Both by Edward S. Ross)

*(Below right) A fragment from a "cross fiber" vein of chrysotile from Seneca, Arizona, shows the delicate fibers into which asbestos can be separated.
(Reo N. Pickens, Jr.)*

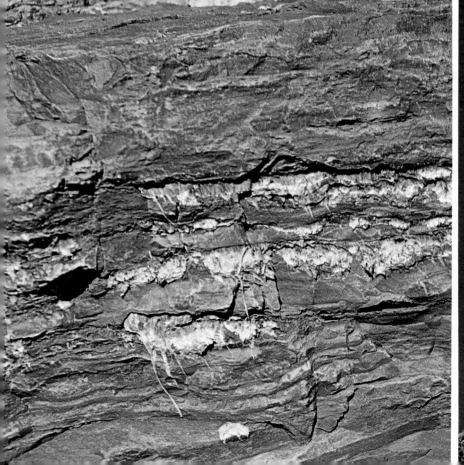

articles are manufactured from asbestos, the kind of product determining the variety of fiber to be used. It was first woven into cloth, as by the ancients, to be made into fireproof garments, and, after many disastrous theatre fires, into theater curtains. But it was not until the use of brake linings in automobiles that asbestos fabrics were produced on a large scale. Although today most brake linings are molded rather than woven, they are still made from asbestos. Other major uses are in asbestos-cement products such as wall board and house shingles. In a new and interesting development, asbestos and a bonding liquid are sprayed onto walls and ceilings, giving not only fire protection but also acoustical and heat-insulating qualities.

Brimstone

Sulfur is an abundant element in the earth's crust and a constituent of many minerals found at or near the earth's surface. Some of it is present in the sulfides, metallic minerals that form the ores of many of the common metals. But there are much greater quantities in the sulfates, particularly gypsum and anhydrite, which have been precipitated in huge quantities from sea water. Although much less widely distributed than these minerals, native sulfur has for centuries been the principal source of the element, its preeminent position being challenged only in recent years.

Under the name of brimstone, meaning the stone that burns, sulfur has been known since ancient times. It is often mentioned in the Bible, and Homer speaks of it both as a "divine and purifying fumigation" and as "pest diverting." After the slaughter of Penelope's suitors, Ulysses, calling for a general purification, cries, "Bring fire that I may burn sulfur, the cure of all ills!" Pliny in his *Historia Naturalis* mentions the role of sulfur in religious ceremonies and in fumigating homes. Because it has a low melting temperature, 235° F, and is completely consumed by burning, it was also a favorite chemical of the alchemists.

Sulfur has several distinctive properties. One of these, the color, is so well known that "sulfur yellow" is often used to describe any similar color; impurities, however, change it to yellow shades of red or green. Sulfur can be ignited with a match, and burns with a blue flame to sulfur dioxide gas. Although different arrangements of the atoms give rise to three crystalline types (polymorphs), the only type common in nature occurs in distinctive orthorhombic crystals. Care should be exercised in handling fine crystals, for sulfur is such a poor heat conductor that the warmth of the hand may damage the specimen by causing the surface to expand and spall from the cooler interior.

Sulfur was known in early times for it is found on the earth's surface in solfatara deposits around volcanos and fumeroles. Among the gases escaping from solfataras are hydrogen sulfide and sulfur dioxide, and by reaction with one another or by oxidation of hydrogen sulfide, these deposit sulfur.

Thus in a geyser issuing from volcanic rocks at Krisuvik, Iceland, first described in 1873, water accompanied by steam, hydrogen sulfide and sulfur dioxide in time deposited a crust of sulfur from two to three feet thick. The strong-smelling vapors of hydrogen sulfide and sulfur dioxide rise even from quiescent volcanos and, as can be seen at Mount Vesuvius, deposit sulfur around the rim. Some extinct volcanos long ago formed large amounts of sulfur

Blue-white crystals of celestite have grown from a base of crystalline sulfur. (Studio Hartmann)

in this manner. Such is the type of deposit in Chile, at an elevation of nineteen thousand feet, where vast tonnages of pure sulfur are mined. Similar deposits are also worked in Italy and Japan.

To ancient peoples volcanic eruptions and the sulfur associated with them were tangible evidence of satanic power, and Hellfire and brimstone were commonly mentioned together. When in the nineteenth century sulfur was found in rocks in Sicily near the volcanic cone of Mount Etna, it was assumed that it came from subterranean vapors. However, we now know that the sulfur had a completely different origin. The Sicilian deposits, found within an area of eighty by thirty-five miles in the central part of the island, are contained in essentially flat-lying sedimentary rocks composed of porous limestone, associated with beds of the calcium sulfates, anhydrite and gypsum. It is now generally agreed that the sulfur derives from the anhydrite and gypsum. Some thought it resulted from inorganic chemical reactions, but more recently it has been attributed to bacterial action. As animals derive energy from the oxidation of carbon, certain bacteria derive their energy from the oxidation of sulfur, leaving the uncombined element as an end product of their life processes.

The mines in Sicily, particularly near Girgenti, have furnished not only most of the fine specimens of crystallized sulfur but also superb specimens of other minerals, particularly crystals of celestite and aragonite.

Sulfur has been mined in Sicily since about A.D. 1250 but it was not until the end of the eighteenth century that it became a significant industry. At that time the Leblanc process for the production of sodium carbonate essential in the manufacture of glass required large quantities of sulfur to make sulfuric acid. The mining operation, however, remained not only primitive but injurious to all living things in the neighborhood of the mines. Women and boys carried out the ore in baskets on their heads, and dumped it into a large hole in the ground where it was ignited. As much as two-thirds of the sulfur was consumed in order to melt the remainder, which was then poured into vats to crystallize. The escaping sulfur dioxide killed the vegetation for miles around. In the nineteenth century a slightly more efficient method was introduced in which about half of the sulfur was recovered. Although small amounts were produced elsewhere, Sicily had virtually a world monopoly until an ingenious process was devised whereby the sulfur was melted in depth and brought to the surface in molten form.

In 1868 oil drillers on the Gulf coast of Louisiana encountered at a depth of six hundred feet a layer of porous limestone, one hundred feet thick, impregnated with sulfur. Below the sulfur-bearing rock was a great but undetermined thickness of gypsum and anhydrite; above it twenty-five feet of limestone overlain by quicksand. The association of sulfur with limestone, anhydrite and gypsum was the same as that in Sicily. The obvious method of recovery was to sink a shaft, mine the rock, hoist it to the surface, and there extract the sulfur. For twenty years, hundreds of thousands of dollars were invested in sinking a shaft, but the engineering skills of the time were unable to cope with the layer of quicksand. Conventional methods of reaching the sulfur were abandoned in 1890 when five men were asphyxiated by hydrogen sulfide gas escaping from the water welling up in the bottom of the shaft. Finally, in 1890, the German-born American chemist, Herman Frasch, patented a process for

(Above) Pyrite crystals from Mexico. In some countries where pyrite is abundant it is used as a source of sulfur. (Floyd R. Getsinger)

(Below) Sulfur in brilliantly faced crystals lines a cavity from the Agrigento Mines, Sicily. (Reo N. Pickens, Jr.)

bringing the sulfur to the surface. The principle was relatively simple. Although water boils at 212° F at sea level, it can, under pressure, be heated to higher temperatures, higher than the melting point of sulfur (235° F). If such superheated water could be forced down, it would melt the sulfur which could then be pumped to the surface. The theory was ridiculed by many. But in December, 1894, after superheated water had poured into the well for twenty-four hours and the outlet valve was opened, a golden brown stream of liquid sulfur poured forth. In a few years the revolutionary mining method changed the United States from a large importer of sulfur to a major exporter and world supplier. Although the Frasch process has since been improved, it still operates on the same basic principles.

Studies of the geology of the sulfur-bearing rocks in Louisiana revealed that they formed part of the cap rock overlying a salt dome. In the course of searching for associated petroleum, over one hundred salt domes have been located on the Gulf coast of Louisiana and Texas. The cap rock of about half of the domes contain sulfur, but in only ten or twelve has it been found in commercially significant quantities. The domes range in area from less than one hundred acres to several square miles. One of the largest, and the most productive is Bowling Dome in Texas; production began there in 1927 and over fifty million tons of sulfur were recovered from it during the next thirty years. The molten sulfur is brought to the surface and run into areas measuring 1000 × 200 feet inclosed by low walls. As the sulfur crystallizes and hardens, the walls are moved continuously upward until a cake forty feet high is formed, containing 500,000 tons of sulfur. Bowling Dome's production was surpassed in 1963 by the sulfur mine at Grande Ecaille salt dome in Louisiana. Sulfur is also being recovered from salt domes on the Isthmus of Tehuantepec, Mexico, and offshore in the shallow waters of the Gulf of Mexico. The recovery of petroleum associated with the domes in the Gulf has received much publicity, but the production of sulfur is equally impressive. The Grande Isle mine, a $30,000,000 investment located seven miles from land, is a monument to the resourcefulness of the sulfur miner.

Sulfur is still used in insecticides and fumigation as it was two thousand years ago. However, it also has so many other important commercial uses that its consumption is sometimes taken as a measure of the industrial vitality of a country. The largest amounts are consumed by the fertilizer industry where, after being converted to sulfuric acid, it is used to produce superphosphate and ammonium sulfate. Smaller quantities are used in the manufacture of steel, aluminum, paper, textiles and pigments.

(Above) A stockpile of sulfur at Basel, Switzerland, ready for shipment. (Pierre A. Pittet)

(Below) Sulfur pumped from underground in molten form at Bowling Dome, Texas, has solidified into blocks forty feet high, each covering about five acres. (Cornelius S. Hurlbut, Jr.)

Appendix

Properties of Some Common Minerals

In the following table are listed the minerals mentioned in this book with their chemical compositions and some of their physical properties.

Name	Chemical Composition	Crystal System	Luster	Specific Gravity	Hardness	Remarks
Actinolite	Ca, Mg, Fe silicate	Mon	Vit.	3.0–3.2	5–6	An amphibole
Albite	$Na(AlSi_3O_8)$	Tric	Vit.	2.62	6	A feldspar
Almandite	$Fe_3Al_2(SiO_4)_3$	Iso	Vit.	4.25	7	A garnet
Amblygonite	$LiAlFPO_4$	Tric	Vit.	3.0–3.1	6	Fusible at 2
Analcime	$Na(AlSi_2O_6)H_2O$	Iso	Vit.	2.27	5–5½	A feldspathoid
Andalusite	Al_2SiO_5	Orth	Vit.	3.16–3.20	7½	Infusible
Andradite	$Ca_3Fe_2(SiO_4)_3$	Iso	Vit.	3.75	7	A garnet
Anglesite	$PbSO_4$	Orth	Ada.	6.2–6.4	3	3 cleav. direc.
Anhydrite	$CaSO_4$	Orth	Vit.	2.89–2.98	3–3½	3 cleav. direc.
Anorthite	$CaAl_2Si_2O_8$	Tric	Vit.	2.76	6	2 cleav.
Apatite	$Ca_5(F, Cl, OH)(PO_4)_3$	Hex	Vit.	3.15–3.20	5	1 poor cleav.
Apophyllite	$Ca_4K(Si_4O_{10})_2 F \cdot 8H_2O$	Tet	Vit.	2.3–2.4	4½–5	1 good cleav.
Aragonite	$CaCO_3$	Orth	Vit.	2.95	3½–4	2 cleav.
Argentite	Ag_2S	Iso	Metal.	7.3	2–2½	Sectile
Arsenopyrite	$FeAsS$	Mon	Metal.	6.07±0.15	5½–6	Silvery
Augite	Ca, Mg, Fe, Al silicate	Mon	Vit.	3.2–3.4	5–6	Black, 2 cleav.
Autunite	$Ca(UO_2)_2(PO_4)_2 \cdot 10-12H_2O$	Tet	Vit.	3.1–3.2	2–2½	Yellow-green
Azurite	$Cu_3(CO_3)_2(OH)_2$	Mon	Vit.	3.77	3½–4	Always blue
Barite	$BaSO_4$	Orth	Vit.	4.5	3–3½	3 cleav. direc.
Bauxite	A mixture of aluminum hydroxides	Amor	Earth.	2.0–2.55	1–3	An earthy rock
Beryl	$Be_3Al_2(Si_6O_{18})$	Hex	Vit.	2.75–2.8	7½–8	Usually green
Biotite	K, Mg, Fe, Al silicate	Mon	Vit.	2.8–3.2	2½–3	Black mica
Borax	$Na_2B_4O_7 \cdot 10H_2O$	Mon	Vit.	1.7±	2–2½	1 cleav.
Bornite	Cu_5FeS_4	Iso	Metal.	5.06–5.08	3	Purple-blue tarnish
Calaverite	$AuTe_2$	Mon	Metal.	9.35	2½	Easily fusible
Calcite	$CaCO_3$	Rho	Vit.	2.71	3	Good cleav.
Carnotite	$K_2(UO_2)_2(VO_4)_2 \cdot nH_2O$	Orth	Vit.	4.1–5.0	Soft	Yellow
Cassiterite	SnO_2	Tet	Ada.	6.8–7.1	6–7	Colorless to black
Celestite	$SrSO_4$	Orth	Vit.	3.95–3.97	3–3½	3 cleav. direc.
Cerargyrite	$AgCl$	Iso	Waxlike	5.5±	2–3	Perfectly sectile
Cerussite	$PbCO_3$	Orth	Ada.	6.55	3–3½	Good cleav.
Chabazite	Ca, Na, Al silicate	Rho	Vit.	2.05–2.15	4–5	Cubelike crystals
Chalcanthite	$CuSO_4 \cdot 5H_2O$	Tric	Vit.	2.12–2.30	2½	Soluble in water
Chalcocite	Cu_2S	Orth	Metal.	5.5–5.8	2½–3	Imperfectly sectile
Chalcopyrite	$CuFeS_2$	Tet	Metal.	4.1–4.3	3½–4	Brittle. Yellow.
Chlorite	Mg, Fe silicate	Mon	Vit.	2.6–2.9	2–2½	1 good cleav.
Chromite	$FeCr_2O_4$	Iso	Metal.	4.6	5½	Luster submetallic

Name	Chemical Composition	Crystal System	Luster	Specific Gravity	Hardness	Remarks
Chrysoberyl	$BeAl_2O_4$	Orth	Vit.	3.65–3.8	8½	Crystals tabular
Chrysocolla	$CuSiO_3 \cdot 2H_2O$	?	Vit.	2.0–2.4	2–4	Bluish-green
Cinnabar	HgS	Rho	Ada.	8.10	2½	Red
Cobaltite	$CoAsS$	Iso	Metal.	6.33	5½	In crystals
Colemanite	$Ca_2B_6O_{11} \cdot 5H_2O$	Mon	Vit.	2.42	4–4½	1 perfect cleav.
Copper	Cu	Iso	Metal.	8.9	2½–3	Malleable
Corundum	Al_2O_3	Rho	Vit.	4.02	9	Various colors
Covellite	CuS	Hex	Metal.	4.6–4.76	1½–2	Blue
Crocoite	$PbCrO_4$	Mon	Ada.	5.9–6.1	2½–3	Orange-red
Cryolite	Na_3AlF_6	Mon	Vit.	2.95–3.0	2½	White
Cuprite	Cu_2O	Iso	Ada.	6.0	3½–4	In red crystals
Datolite	$CaB(SiO_4)(OH)$	Mon	Vit.	2.8–3.0	5–5½	Usually in crystals
Diamond	C	Iso	Ada.	3.5	10	In crystals
Diopside	$CaMg(Si_2O_6)$	Mon	Vit.	3.2–3.3	5–6	A pyroxene
Dolomite	$CaMg(CO_3)_2$	Rho	Vit.	2.85	3½–4	Good cleav.
Enargite	Cu_3AsS_4	Orth	Metal.	4.43–4.45	3	Silvery
Epidote	Ca, Fe, Al silicate	Mon	Vit.	3.35–3.45	6–7	Green
Fluorite	CaF_2	Iso	Vit.	3.18	4	Octahedral cleav.
Franklinite	Fe, Zn, Mn oxide	Iso	Metal.	5.15	6	Black
Gahnite	$ZnAl_2O_4$	Iso	Vit.	4.55	7½–8	Green octahedrons
Galena	PbS	Iso	Metal.	7.4–7.6	2½	Cubic cleav.
Garnet	$A_3'' B_2''' (SiO_4)_3$	Iso	Vit.	3.5–4.3	6½–7½	In crystals
Garnierite	$(Ni, Mg)SiO_3 \cdot nH_2O$	Amor	Vit.	2.2–2.8	2–3	Green
Goethite	$HFeO_2$	Orth	Ada.	4.37	5–5½	1 cleav.
Gold	Au	Iso	Metal.	15.0–19.3	2½–3	Yellow. Soft.
Graphite	C	Hex	Metal.	2.3	1–2	Black. Platy.
Grossularite	$Ca_3Al_2(SiO_4)_3$	Iso	Vit.	3.53	6½	A garnet
Gypsum	$CaSO_4 \cdot 2H_2O$	Mon	Vit.	2.32	2	White
Halite	$NaCl$	Iso	Vit.	2.16	2½	Cubic cleav., Salty.
Hematite	Fe_2O_3	Rho	Metal.	5.26	5½–6½	Red powder
Hemimorphite	$Zn_4(Si_2O_7)(OH)_2 \cdot H_2O$	Orth	Vit.	3.4–3.5	4½–5	Cleav.
Heulandite	$Ca(Al_2Si_7O_{18}) \cdot 6H_2O$	Mon	Vit.	2.18–2.20	3½–4	1 cleav.
Hornblende	Ca, Mg, Fe, Al silicate	Mon	Vit.	3.2	5–6	Cleav.
Idocrase	Ca, Mg, Fe, Al silicate	Tet	Vit.	3.35–3.45	6½	Prismatic crystals
Ilmenite	$FeTiO_3$	Rho	Metal.	4.7	5½–6	Slightly magnetic
Iron	Fe	Iso	Metal.	7.3–7.9	4½	Magnetic
Jadeite	$NaAl(Si_2O_6)$	Mon	Vit.	3.3–3.5	6½–7	Green. Compact.
Kaolinite	$Al_4(Si_4O_{10})(OH)_8$	Mon	Vit.	2.6–2.65	2–2½	Earthy
Kernite	$Na_2B_4O_7 \cdot 4H_2O$	Mon	Vit.	1.95	3	2 cleav.
Kyanite	Al_2SiO_5	Tric	Vit.	3.56–3.66	5–7	Blue. Bladed.
Lazurite	Na, Ca, Al silicate	Iso	Vit.	2.4–2.45	5–5½	Pyrite associated
Lepidolite	K, Li, Al silicate	Mon	Vit.	2.8–3.0	2½–4	A mica
Leucite	$K(AlSi_2O_6)$		Vit.	2.45–2.50	5½–6	In crystals
Limonite	$FeO(OH) \cdot nH_2O$	Amor	Vit.	3.6–4.0	5–5½	Yellow-brown
Magnesite	$MgCO_3$	Rho	Vit.	3.0–3.2	3½–5	Commonly massive
Magnetite	Fe_3O_4	Iso	Metal.	5.18	6	Strongly magnetic

Name	Chemical Composition	Crystal System	Luster	Specific Gravity	Hardness	Remarks
Malachite	$Cu_2CO_3(OH)_2$	Mon	Vit.	3.9–4.03	3½–4	Green
Marcasite	FeS_2	Orth	Metal.	4.89	6–6½	In crystals
Microcline	$K(AlSi_3O_8)$	Tric	Vit.	2.54–2.57	6	A feldspar
Millerite	NiS	Rho	Metal.	5.5±0.2	3–3½	Capillary crystals
Mimetite	$Pb_5Cl(AsO_4)_3$	Hex	Ada.	7.0–7.2	3½	Like pyromorphite
Molybdenite	MoS_2	Hex	Metal.	4.62–4.73	1–1½	Black. Platy
Monazite	$(Ce, La, Y, Th)PO_4$	Mon	Vit.	5.0–5.3	5–5½	Small xtals.
Muscovite	$KAl_2(AlSi_3O_{10})(OH)_2$	Mon	Vit.	2.76–3.1	2–2½	1 perfect cleav.
Natrolite	$Na_2(Al_2Si_3O_{10}) \cdot 2H_2O$	Mon	Vit.	2.25	5–5½	1 perfect cleav.
Nepheline	$(Na, K)(AlSiO_4)$	Hex	Vit.	2.55–2.65	5½–6	Greasy appear.
Niccolite	$NiAs$	Hex	Metal.	7.78	5–5½	Copper-red
Olivine	$(Mg, Fe)_2SiO_4$	Orth	Vit.	3.27–4.37	6½–7	Green rock min.
Opal	$SiO_2 \cdot nH_2O$	Amor	Vit.	1.9–2.2	5–6	Non-crystal.
Orpiment	As_2S_3	Mon	Vit.	3.49	1½–2	1 cleav. Yellow.
Orthoclase	$K(AlSi_3O_8)$	Mon	Vit.	2.57	6	A feldspar
Pectolite	$Ca_2NaH(SiO_3)_3$	Tric	Vit.	2.7–2.8	5	Crystals acicular
Pentlandite	$(Fe, Ni)_9S_8$	Iso	Metal.	4.6–5.0	3½–4	In pyrrhotite
Phlogopite	$KMg_3(AlSi_3O_{10})(OH)_2$	Mon	Vit.	2.86	2½–3	Brown mica
Platinum	Pt	Iso	Metal.	14–19	4–4½	Silvery
Prehnite	$Ca_2Al_2(Si_3O_{10})(OH)_2$	Orth	Vit.	2.8–2.95	6–6½	Tabular crystals
Pyrargyrite	Ag_3SbS_3	Rho	Ada.	5.85	2½	Dark ruby silver
Pyrite	FeS_2	Iso	Metal.	5.02	6–6½	Crystals striated
Pyrolusite	MnO_2	Tet	Metal.	4.75	1–2	Sooty
Pyromorphite	$Pb_5Cl(PO_4)_3$	Hex	Ada.	6.5–7.1	3½–4	Adamantine luster
Pyrope	$Mg_3Al_2(SiO_4)_3$	Iso	Vit.	3.51	7	A garnet
Pyrrhotite	$Fe_{1-x}S$	Hex	Metal.	4.58–4.65	4	Magnetic
Quartz	SiO_2	Rho	Vit.	2.65	7	No cleavage
Realgar	AsS	Mon	Res.	3.48	1½–2	1 cleav. Red.
Rhodochrosite	$MnCO_3$	Rho	Vit.	3.45–3.6	3½–4½	Pink
Rhodonite	$Mn(SiO_3)$	Tric	Vit.	3.58–3.70	5½–6	Pink
Rutile	TiO_2	Tet	Ada.	4.18–4.25	6–6½	Small cryst.
Scheelite	$CaWO_4$	Tet	Ada.	5.9–6.1	4½–5	4 cleav. direc.
Serpentine	$Mg_6(Si_4O_{10})(OH)_8$	Mon	Vit.	2.2	2–5	Green to yellow
Siderite	$FeCO_3$	Rho	Vit.	3.83–3.88	3½–4	Brown
Sillimanite	Al_2SiO_5	Orth	Vit.	3.23	6–7	1 cleav.
Silver	Ag	Iso	Metal.	10.5	2½–3	White, malleable
Smithsonite	$ZnCO_3$	Rho	Vit.	4.35–4.40	5	Reniform
Sodalite	$Na_4(AlSiO_4)_3Cl$	Iso	Vit.	2.15–2.3	5½–6	Usually blue
Spessartite	$Mn_3Al_2(SiO_4)_3$	Iso	Vit.	4.18	7	A garnet
Sphalerite	ZnS	Iso	Res.	3.9–4.1	3½–4	6 cleav. direc.
Sphene	$CaTiO(SiO_4)$	Mon	Ada.	3.40–3.55	5–5½	Wedge-shaped Xls
Spinel	$MgAl_2O_4$	Iso	Vit.	3.6–4.0	8	In octahedrons
Spodumene	$LiAl(Si_2O_6)$	Mon	Vit.	3.15–3.20	6½–7	Good cleav.
Staurolite	$Fe_2Al_9O_7(SiO_4)_4(OH)$	Orth	Vit.	3.65–3.75	7–7½	In cruciform twins
Stibnite	Sb_2S_3	Orth	Metal.	4.52–4.62	2	Perfect cleav.
Stilbite	$Ca(Al_2Si_7O_{18}) \cdot 7H_2O$	Mon	Vit.	2.1–2.2	3½–4	Sheaflike agg.

Name	Chemical Composition	Crystal System	Luster	Specific Gravity	Hardness	Remarks
Strontianite	$SrCO_3$	Orth	Vit.	3.7	$3\frac{1}{2}-4$	Efferv. in HCl
Sulfur	S	Orth	Res.	2.05–2.09	$1\frac{1}{2}-2\frac{1}{2}$	Burns
Sylvanite	$(Au, Ag)Te_2$	Mon	Metal.	8.0–8.2	$1\frac{1}{2}-2$	Perfect cleav.
Talc	$Mg_3(Si_4O_{10})(OH)_2$	Mon	Vit.	2.7–2.8	1	Greasy feel
Tennantite	$(Cu, Fe, Zn, Ag)_{12}As_4S_{13}$	Iso	Metal.	4.6–5.1	$3-4\frac{1}{2}$	In tetrahedrons
Tetrahedrite	$(Cu, Fe, Zn, Ag)_{12}Sb_4S_{13}$	Iso	Metal.	4.6–5.1	$3-4\frac{1}{2}$	In tetrahedrons
Topaz	$Al_2(SiO_4)(F, OH)_2$	Orth	Vit.	3.4–3.6	8	Perfect cleav.
Tourmaline	$XY_3Al_6(BO_3)_3(Si_6O_{18})(OH)_4$	Rho	Vit.	3.0–3.25	$7-7\frac{1}{2}$	Trigonal section
Tremolite	$Ca_2Mg_5(Si_8O_{22})(OH)_2$	Mon	Vit.	3.0–3.3	5–6	Cleav.
Turquoise	Cu, Al phosphate (hydrous)	Tric	Vit.	2.6–2.8	6	Blue-green
Uraninite	UO_2	Iso	Res.	9.0–9.7	$5\frac{1}{2}$	Pitchy luster
Uvarovite	$Ca_3Cr_2(SiO_4)_3$	Iso	Vit.	3.45	$7\frac{1}{2}$	Green garnet
Vanadinite	$Pb_5Cl(VO_4)_3$	Hex	Vit.	6.7–7.1	3	Luster resinous
Willemite	Zn_2SiO_4	Rho	Vit.	3.9–4.2	$5\frac{1}{2}$	From Franklin, N.J.
Witherite	$BaCO_3$	Orth	Vit.	4.3	$3\frac{1}{2}$	Efferv. in HCl
Wollastonite	$Ca(SiO_3)$	Tric	Vit.	2.8–2.9	$5-5\frac{1}{2}$	2 cleav.
Wulfenite	$PbMoO_4$	Tet	Ada.	$6.8\pm$	3	Orange-red
Zincite	ZnO	Hex	Vit.	5.68	$4-4\frac{1}{2}$	At Franklin, N.J.
Zircon	$ZiSiO_4$	Tet	Ada.	4.68	$7\frac{1}{2}$	In small crystals

Index

Asterisks indicate pages containing illustrations

Accessory minerals, 60–66
Achroite, 82
Actinolite, 257, 284, 293
Agate, 36, 94*, 228*, 229, 233*, 234*, 237*, 238*, 242, 245, 250, 252*
 moss, 245
Agricola, Georgius, 11, 98, 152, 189
Ajanta caves, 94*
Ajmer, India, 223
Alabaster, 127–128, 129
Alamagordo, New Mexico, 129
Albertus Magnus, 11
Albite, 47, 77, 82, 87, 90*, 293
Alexander the Great, 146
Alexandrite, 85, 222
Alicante, Colorado, 254
Alkali Lake (California), 137
Allanite, 80
Alma, Colorado, 254
Almaden, Spain, 163
Almandite, 69*, 258, 259, 260, 293
Alsace-Lorraine, 205
Aluminum, 34, 41, 45, 88, 95, 159, 160, 178, 205, 258
Amargosa Borax Works, 138
Amazonite, 36, 70*, 75*, 77
Amblygonite, 73, 79, 80, 293
Amelia Court House, Virginia, 260
American River, 149
Amethyst, 20, 94*, 99, 220, 221, 228, 236, 239
Amosite, 284
Amphibole, 41, 45, 46, 245, 257, 284
Analcime, 95, 97, 98, 293
Andalusite, 54, 66, 67–68, 293
Andesine, 47, 48
Andesite, 48
Andes Mountains, 48
Andradite, 59, 256*, 258, 260, 293

Angle Mounds (Indiana), 115
Anglesite, 175, 293
Anhalt, Germany, 123
Anhydrite, 52, 121, 122, 124, 127, 129, 287, 288, 293
Annaberg, Saxony, 151
Anorthite, 47, 48, 293
Anorthosite, 154*
Anthophyllite, 284
Antimony, 159, 160, 163, 164, 168
Antlerite, 167, 175
Antrim County, Ireland, 98, 100
Apatite, 14, 21, 47, 48, 60, 62*, 65–66, 79, 88, 136*, 202, 293
Apophyllite, 93*, 99, 156, 293
Aquamarine, 85, 89*, 223
Aragonite, 12, 104, 108*, 109, 206*, 288, 293
Argentite, 167, 293
Arsenic, 160, 163–164, 168, 185, 277
Arsenides, 151, 152
Arsenopyrite, 163, 293
Asbestos, 282*, 283–284, 285*, 287
Ash, volcanic, 45, 50, 100
Ashcroftine, 97
Asparagus stone, 62*
Asterism, 79, 219
Atacama desert, 175
Atomic energy, minerals and, 264–271
Auburn, Maine, 66
Augite, 41, 43*, 65*, 249, 257, 293
Aurichalcite, 174*
Autunite, 79, 265*, 270, 293
Aventurine, 77
Avery Island (Louisiana), 124
Azurite, 18*, 19, 155, 170*, 171, 172, 278, 293

Babingtonite, 93*, 99
Bad Lands (South Dakota), 106
Baikal, Lake (Siberia), 250
Bancroft, Ontario, 250
Bangkok, Thailand, 220
Barite, 16, 18*, 111, 113*, 114*, 115*, 116, 117, 124, 196, 277, 293
Barium, 41, 101
Barringer Crater, 193

Barringer, D.M., 194
Basalt, 42, 43*, 48–49, 92, 95, 98, 100
Base exchange, 100–101
Battambang, Cambodia, 220
Bauxite, 88, 178, 179*, 205, 293
"Baveno twin," 46*
Bergmann (Swedish chemist), 231
Berthier (French chemist), 182
Bertrandite, 88
Beryl, 14, 22, 73, 74, 79, 80, 85, 87, 89*, 222–223, 293
 golden, 85, 89*
Beryllium, 72, 74, 77, 79, 87–88
Berzelius, Jons Jakob, 80, 231
Bessemer process, 190
Bikita, Southern Rhodesia, 91, 186
Bingham Canyon (Utah), 176*
Biot, 231
Biotite, 41, 42, 46, 47, 48, 54, 57, 74, 78, 88, 293
Bisbee, Arizona, 172
Bismuth, 151, 152, 270
Black Hills (South Dakota), 14, 72, 73, 240
Blast furnace, 202
Bloodstone, 229, 245
"Blue John," 115
Boehmite, 178
"Bog ore," 189, 205
Boracite, 124, 132
Borates, 132–141
Borax, 132–133, 134, 137–138, 139*, 140, 293
Borax Lake, 137
Boric acid, 134
Bornite, 165*, 167, 277, 278, 293
Boron, 72, 79, 132, 133, 134, 137–141, 275, 276
Boron, California, 140
Boron carbide, 133
Botallack mine, 280
Bou Beker, Morocco, 104
Bowling Dome (Texas), 291
Boyle, Robert, 229
Bradbury Mountain, Maine, 78
Branchville, Connecticut, 79
Brandon, England, 30
Brannan, Samuel, 143
Brass, 164
Brazilianite, 80, 86, 89*
Breccia, 45, 50

296

Brewsterite, 97
Brimstone, 287–288
Broken Hill, Australia, 177
Bromyrite, 177
Bronze, 272
Bronze Age, 272
Bruce, Archibald, 182
Budge, E.A.W., 228
Burma, 57, 67, 219
Bushveld Igneous Complex, 158, 259
Butte, Montana, 177
Bytownite, 47

Cadmium, 164, 176
Calaverite, 149, 293
Calcareous tufa, 103
Calcite, 10*, 20, 21, 22, 39*, 52, 57, 95, 98, 99, 101*, 103, 104, 105*, 106, 107*, 109, 117, 124, 127, 148*, 156, 170*, 174*, 180*, 183*, 185, 186, 198, 200*, 201, 223, 249, 251*, 277, 293
Calcium, 41, 45, 46, 59, 95, 101, 102, 160
Calcium carbonate, 103, 104, 110*, 118, 121
Calcium-larsenite, 186
Calcium phosphate, 65
Calcium sulfate, 118, 121
Caleraine, Canada, 284
Calico, California, 140
Californite, 257, 258
Campiglia Marittima, Italy, 253
Campione, Mount, 66
Canadian Shield, 72
Canyon Diablo, 193
Cape Blomidon, Nova Scotia, 100
Cape Province, South Africa, 64
Capillitas, Catamarca, Argentina, 254
Carbon, 41
Carbonado, 211
Carbuncle, 259
Carlsbad, Czechoslovakia, 103
Carlsbad, New Mexico, 130
Carnelian, 36, 228, 242
Carnotite, 266*, 270, 271, 278, 293
Caspian Sea, 121, 131
Cassiterite, 20, 274*, 275, 277, 279, 280, 293

Cat's eye, 85
Cave-in-Rock (Illinois), 115
Celestite, 111, 116–117, 124, 184*, 286*, 288, 293
Ceramics, 33–36
Cerargyrite, 151, 177, 293
Cerro Blanco (Argentina), 71, 74
Cerro de Pasco, Peru, 177
Cerussite, 20, 113*, 114*, 173*, 174*, 175, 175*, 293
Cesium, 79, 101
Ceylon, 67, 220
Chabazite, 95, 97, 98, 101*, 293
Chaco Canyon, 253
Chalcanthite, 175, 293
Chalcedony, 36, 229, 234*, 240, 242, 245, 253
Chalcocite, 167, 172, 175, 278, 293
Chalcopyrite, 104, 107*, 146, 158, 162*, 165*, 167, 201, 277, 278, 293
Chalcotrichite, 166*
Chalk, 245
Champlain, Samuel de, 155
Chatham, Carroll F., 224
Chert, 245
Chervetite, 266*
Cheshire, England, 123
Chiastolite, 68
Chihuahua, Mexico, 109
Chlorine, 41
Chlorite, 20, 54, 57, 69*, 99, 240, 245, 277, 293
Chondrodite, 57
Chromite, 154*, 158, 160, 293
Chromium, 41, 158, 160, 219, 258
Chrysoberyl, 79, 80, 85, 86, 90*, 222, 294
Chrysocolla, 36, 170*, 294
Chrysolite, 260
Chrysoprase, 36, 245
Chrysotile, 284, 285*
Chuquicamata, Chile, 175
"Cinder" cones, 49
Cinnabar, 20, 163, 165*, 294
Cinnamon stone, 259
Citrine, 86, 236, 239
Clay, 32–36, 52, 57, 82, 228, 276, 277, 281*
Clear Lake (California), 137
Cleavage, 21–22

Cleavelandite, 77, 82
Climax, Colorado, 178
Clinohedrite, 186
Clinton iron ore beds (Alabama), 204
Coal, 12, 35
Cobalt, 151, 152, 153, 160, 168, 194, 195, 264, 269, 270, 277
Cobalt, Ontario, 153
Cobalt oxide, 35
Cobaltite, 151, 294
Coconimo Plateau, 193
Colemanite, 132, 133, 138, 140, 294
Collins Hill (Connecticut), 73
Coloma, California, 149
Colombia, 223
Color, 19–20
Colorado Plateau, 271
Colored stones, 249–263
Columbia Plateau, 42, 45, 49, 92
Columbite, 91
Columbite-tantalite, 80
Columbium, 91
Conchoidal fracture, 29
Conglomerate, 51
Consolidated Diamond Mine, 210*, 215
Copper, 32, 36, 59, 99, 142, 143, 148*, 150, 151, 153–156, 157, 158, 159, 160, 164, 165*, 167, 171, 173*, 175, 176*, 177, 178, 187, 207, 250, 264, 269, 270, 272, 276, 277, 278, 279, 294
Córdoba, Argentina, 71
Corinth, Greece, 168*
Cornwall, England, 86, 104, 272–280
Coro-Coro, Bolivia, 155
Cortez, Hernando, 146, 223, 258
Corundum, 21, 57, 59, 217*, 218*, 219, 220–222, 249, 258, 294
Corundum Hill (North Carolina), 220
Covellite, 167, 294
Cripple Creek, Colorado, 149
Cristobalite, 248
Crocidolite, 282*, 284
Crocoite, 23*, 294
Cronstedt, Baron, 97, 168
Crust, Earth's, 38, 41, 45, 160
Cryolite, 88, 90*, 178, 294

"Crystal Cave," 117
"Crystal faces," 16
Crystal systems, 14–16
Crystals, 8, 11–12, 14–16, 19, 42, 45
Cubic crystal system, 15
Cullinan diamond, 215, 221*
Cullinan, Thomas, 214
Cumberland, England, 104, 111, 116, 196
Cummington, Massachusetts, 253–254
Cuprite, 148*, 155, 166*, 171, 278, 294
Curies, the, 152, 232, 269
Curtis, Thomas, 280
Cymophane, 85

Dadiardite, 97
Danburite, 124
Datolite, 98, 99, 148*, 156, 294
Dead Sea, 131
Death Valley, 138
DeBeers Consolidated Mines, Limited, 213
Deccan traps, 49, 100
Deevey, Edward S., Jr., 29
Demantoid, 260
de Marignac (Swiss chemist), 80
Density, 25
Derbyshire, England, 111, 115
Desmine, 98
Devil's Postpile, 95
Dewindite, 266*
Diabase, 100
Diamond, 21, 25, 186, 208, 211–216, 217*, 219, 221*, 294
Diaspore, 178
Diatomite, 241*, 242*, 247*, 248
Dichroism, 85
Ding Dong mine, 275
Diopside, 56*, 59, 294
Diorite, 48
Dioscorides, 195
Dolomite, 52, 57, 59, 104, 107*, 111, 124, 201, 231*, 294
Don Basin, Russia, 145
Dunite, 49, 157, 158, 260
Dun Mountains (New Zealand), 49
Durango, Mexico, 66
Durham, England, 104

Earth, the:
  core of, 38, 142
  crust of, 38, 41, 45, 160
  discontinuities of, 37, 38
  mantle of, 38
  origin of, 37, 38
Earthenware, 32–36
Edingtonite, 97
Egypt, Ancient, minerals, in, 36
Eisenglas, 91
Ekaterinburg (Ural Mountains), 223
Elba, 196
Elbaite, 82
Electrolytic process, 88
Elements, native, 142, 159–160
  in earth's crust, 38, 41
Emerald, 20, 208, 218*, 222–224, 225*, 260
Emerson, Ralph Waldo, 78
Emery, 222
Enargite, 167, 294
Epidote, 56*, 59, 98, 99, 156, 294
Epistilbite, 97
Erionite, 95, 97
Erzberg, Styria, 206
Erzgebirge "ore mountains," 151, 269, 270
Eskefiord, Iceland, 106
Etna, Mount, 288
Etta Mine, 14, 73
Eucryptite, 80, 91, 186
Evaporites, 118
Extrusive rocks, 42, 45, 48

Fairfield, Utah, 253
"Fairy crosses," 66
"Fairy stones," 66
Faujasite, 97, 102
Feldspar, 14, 21, 22, 36, 41, 42, 43*, 46, 46*, 47–48, 49, 50, 54, 61*, 62*, 66, 70*, 71, 73, 75*, 77, 78, 79, 82, 87, 89*, 90*, 98, 124, 178, 275, 276, 277
  plagioclase, 47–48, 50*, 74, 77
  potash, 47, 48, 74, 77, 87
Feldspathoids, 50
Fergusonite, 80
Ferrites, 195
Fibrolite, 67
Fingal's Cave, 100
Fleurus, Algeria, 129

Flint, 27*, 29, 30, 30*, 31, 33*, 82, 227, 245
Flos ferri, 109
Flotation, 177, 178
Fluorescence, 112, 185–186
Fluorine, 41, 72, 77, 79, 88, 115, 275, 276
Fluorite, 9*, 16, 17*, 21, 36, 88, 111–112, 113*, 115–116, 184*, 185, 196, 276, 277, 294
Foliation, 54
Fontainebleau, France, 106
"Fool's gold," 146
Fowler, Samuel, 182
Franklin, New Jersey, 57, 181–186
Franklinite, 180*, 181, 182, 185, 294
Frasch, Herman, 288
Frasch process, 288, 291
Freiberg, Germany, 104, 151, 177

Gabbro, 48
Gadolin (Finnish chemist), 80
Gadolinite, 80
Gahnite, 294
Galena, 10*, 36, 88, 90*, 104, 117, 164, 167, 171*, 173*, 175, 177, 201, 256*, 277, 294
Gallium, 159
Garnet, 36, 54, 57, 60, 66, 69*, 220, 256*, 258–260, 294
Garnierite, 294
Geiger counter, 268–269
Geyserite, 248
Ghost crystal, 240
Giant's Causeway, 53*, 95, 100
Gibbs, Reginald E., 232
Gibbsite, 178
Gillette quarry (Connecticut), 82
Girgenti, Sicily, 109, 288
Gismondine, 97
Glauberite, 141, 247
Gmelinite, 97
Gneiss, 54, 66, 71, 72, 98
Goethite, 52, 124, 190, 192*, 193, 197–198, 204, 205, 278, 294
Golconda, India, 212
Gold, 28*, 32, 32*, 36, 38, 142, 143–150, 154, 157, 160, 163, 207, 264, 270, 294

Goniometer, optical, 12
Gonnardite, 97
Gore Mountain (Adirondacks), 260
Gossans, 149, 198, 278
Grand Canyon, 204
Grande Ecaille salt dome (Louisiana), 291
Grande Isle mine, 291
Granite, 40*, 42, 46, 46*, 47–48, 54, 58*, 59, 71, 72, 74, 91, 146, 158, 269, 275, 276, 277
  graphic, 74, 78, 81*
Granodiorite, 48
Graphic granite, 74, 78, 81*
Graphite, 21, 22, 25, 57, 59, 211, 294
Great Bear Lake, 153, 271
Great Salt Lake, Utah, 125, 132
Greenland, 88
"Greisening," 276
"Grog," 35
Grossularite, 59, 256*, 258, 259, 294
Groton, New Hampshire, 79
Guanajuato, Mexico, 177
Gummite, 265*
Gypsum, 14, 21, 52, 109, 111, 116*, 119*, 120*, 121, 122, 126*, 127*, 127–129, 287, 288, 294

Habachtal, Austria, 223
Haddan Neck, Connecticut, 82
Hafnium, 63
Hagendorf, Bavaria, 79
Halite, 21, 120*, 121, 122–126, 127, 129–130, 140, 294
Halliburton-Bancroft district, Ontario, 14
Hamlin, Augustus, 82
Hamlin, Elisha L., 82
Hanchinhama Pool, 137
Hardness, 21
Harmony Borax Works, 138
Harmotome, 97
Hauerite, 124
Haüy, René Just, 12, 15, 109, 230, 235
Haüynite, 50
Hawaiian Islands, 49
Hearne, Samuel, 153
Heliodor, 85

Heliotrope, 229, 245
Hell's Canyon, 92
Helvite, 88
Hematite, 18*, 20, 36, 52, 60, 99, 124, 166*, 187–188, 188*, 190, 191*, 193, 195–197, 200*, 204–205, 240, 245, 294
Hemimorphite, 172*, 176, 179*, 294
"Heraclion stone," 194
Herschel, John, 231
Hessonite, 259
Heulandite, 97, 98, 294
Hexagonal crystal system, 15
Hidden, W.E., 85
Hiddenite, 83*, 85
Hilgardite, 124
Hoffer, Hubert, 134
Holmes, Ezekiel, 82
Hoover, Herbert, 11, 189
"Hopewell Furnace" (Pennsylvania), 190
"Hopper crystals," 125
Hornblende, 41, 47, 48, 54, 66, 249, 257, 260, 294
"Hornfels," 59
Houghton, Douglass, 155–156
Huddleston, J.W., 216
Hull-Rust-Sellers mine, 191*
Humboldt County, Nevada, 247
Hunan, China, 164
Huyghens, Christiaan, 230, 235
Hyacinth, 61*, 63
Hydraulicing, 276
Hydrogen, 41

Iceland spar, 106
Idaho batholith, 42
Idocrase, 44*, 59, 254, 257, 258, 294
Igneous rocks, 42–50, 57, 60, 71, 72, 158
  accessory minerals of, 60
Ilmenite, 47, 48, 60, 64, 193, 202, 203, 294
Index of refraction, 20, 21
Indicolite, 82
Indium, 164
Intrusive rocks, 42, 45, 47, 48
Inyo Mountains (California), 67
Iodyrite, 177
Ion exchange, 101
Iron, 38, 41, 45, 58, 77, 88, 103, 142, 151, 157, 158, 159, 160, 163, 164, 170*, 178, 182, 187–207, 257, 258, 259, 272, 276, 294
Iron oxide, 118, 121
Isidium, 157
Isinglass, 91
Isle Haute, Nova Scotia, 100
Isle, Rome de l', 230, 235
Isle Royale (Lake Superior), 154, 156
Isometric crystal system, 15
Itacolumite, 58
Itrongay, Madagascar, 77
Ivigtut, Greenland, 88

Jacinth, 63
Jade, 36, 245, 252*, 257–258
Jadeite, 257–258, 294
Jasper, 36, 228, 245–246
Jefferson Island (Louisiana), 124
Jenkin, A.K.H., 279, 280
Joachimsthal, Bohemia, 151, 269, 270
Johannesburg, South Africa, 38, 214, 250
Jur tribe, 188–189

Kalgoorlie, Australia, 149
Kaolinite, 34, 276–277, 294
Kaolinization, 276
Karabib, Southwest Africa, 91
Kara Bogaz Gulf, 121
Kashmir, 220
Kasolite, 266*
Katanga, Congo, 172
Kazakhstan, Russia, 68
Kelly, New Mexico, 176
Kern County, California, 140
Kernite, 132, 133, 139*, 140, 294
Keweenaw Peninsula, 98, 99, 154, 155, 156
Khibina tundra, 65
Kidney ore, 197
Kimberley, South Africa, 209*
Kimberlite, 212, 217*
Kiruna, Sweden, 48, 202
Klaproth, M.H., 63, 269
Kokeha Valley, 250
Kola Peninsula (Russia), 50, 65
Kongsberg, Norway, 99, 151
Kötschubeite, 69*
Kristiansand, Norway, 14

Krisuvik, Iceland, 287
Krystallos, 228
Kunz, G. F., 85
Kunzite, 84*, 85
Kyanite, 54, 56*, 57, 66, 67, 67*, 68, 294

La Bine, Gilbert, 153, 270
Labrador, 77
Labradorite, 47, 48, 75*, 77
Laco iron ore deposit, 199*
Lampoc, California, 248
Lancashire, England, 104
Lancaster, Pennsylvania, 190
Land's End, Cornwall, England, 273*, 275
Langban, Sweden, 186, 253
Lapis lazuli, 19, 36, 249–250, 251*
Lapsa Buru, India, 68
Larderel, Francesco, 134
Laterites, 205
Laumonite, 97
Laurionite, 177
Laurium, Greece, 176, 177
Lava flows, 42, 45, 48–49, 50, 53*, 55*, 156
Lazurite, 50, 249–250, 251*, 294
Lead, 59, 88, 104, 111, 159, 160, 163, 164, 175, 177, 185, 201, 267, 272, 276
Leakey, L. S. B., 26, 29
Leblanc process, 288
Lepidolite, 39*, 78, 79, 80, 88, 91, 294
Leucite, 49*, 50, 294
Leucite syenite, 50
Levyne, 97
Lewiston, Utah, 253
Lick Springs, California, 137
Lightning Ridge field (Australia), 247
Limestone, 38, 52, 57, 59, 65, 103, 104, 109, 111, 117, 121, 176, 198, 201, 202, 204, 219, 220, 223, 245, 250, 253, 257, 288
  Franklin, 57
Limonite, 23*, 146, 171, 179*, 190, 197–198, 205, 294
Lithia mica, 91
Lithium, 72, 74, 77, 79, 82, 87, 91, 276
Lithium silicate, 14

Lithium stearate, 91
Lodestone, 194
Los Cerillos Mountains (New Mexico), 253
Luminescence, 111–112, 185–186
Luster, 20–21
Luxembourg, 205

Madagascar, 77, 85
Maghemite, 194
Magma, 42, 45, 48, 49, 50, 59, 60, 72, 77, 79, 158, 160, 163
Magmatic differentiation, 45, 46
Magnesite, 52, 124, 294
Magnesium, 41, 45, 58, 59, 159, 160, 195, 259
Magnesium chloride, 118, 121
Magnesium sulfate, 118, 121
Magnetite, 47, 48, 60, 182, 190, 193, 194–195, 197, 199*, 202–203, 204, 222, 294
Magnetogorsk (Ural Mountains), 197
Malachite, 18*, 19, 36, 155, 166*, 169*, 170*, 171–172, 278, 295
Mammoth Cave (Kentucky), 105*
Mammoth Hot Springs, 103, 110*
Manganese, 41, 160, 182, 186, 195, 254
Manganite, 254*
Mantle, Earth's, 38
Maravilla Mine (Mexico), 109
Marble, 57, 59, 219
Marcasite, 104, 193, 200*, 201, 207, 295
Marshall, James, 143
Martite, 197
Matura, Ceylon, 63
Mercury, 144, 159, 160, 163
Merensky, Hans, 158
Merritt, Cassius, 190
Mesabi Range, 190, 204
Mesolite, 92, 97
"Metakaolin," 34
Metamorphic minerals, 66–68
Metamorphic rocks, 52–59, 66
Metamorphism, 54, 59
Meteor Crater, 193, 194
Meteorites, 193–194, 195*, 196*, 262*, 263

Mica, 22, 39*, 41, 42, 47, 54, 57, 66, 71, 73, 74, 77, 78–79, 82, 87, 88, 91, 146, 275, 276, 277
  lithia, 91
Mica, Mt., 82
Microcline, 47, 70*, 75*, 77, 295
Microlite, 80
Miller, W. G., 153
Millerite, 166*, 168, 295
Mimetite, 175, 295
Mineralogy, 8, 11
Minas Gerais, Brazil, 78, 85, 86, 196, 236, 240
Mine Hill, New Jersey, 182
Minerals:
  accessory, 60–66
  atomic energy and, 264–271
  borate, 132–141
  Cornwall, England, and, 272–280
  crystals of, 8, 11–12, 14–16, 18
  defined, 12
  early use of, 26–36
  fluorescent, 181–186
  metamorphic, 66–68
  ornamental, 249–263
  physical properties of, 19–25
  primary, 163–168
  rock-making, 41
  sea, 118–130
  secondary, 163, 168–177
  specific gravity of, 22
Minette iron ore, 205
Mirabilite, 141
Mogaung, Upper Burma, 257
Mogok, Burma, 219, 220
"Moho," 38
Mohorovičič (Yugoslav scientist), 38
Mohorovičič discontinuity, 38
Mohs, Friedrich, 21
Mohs Scale of Hardness, 21
Mojave Desert, 140
Molecular sieves, 102
Molybdenite, 178, 295
Molybdenum, 159, 160, 178
Monazite, 60, 64, 295
Monoclinic crystal system, 15
Montezuma, 223, 258
Monzonite, 48
Moonstone, 77
Mordenite, 97
Morganite, 85, 223

Mosander, 80
Moss agate, 245
Mother Lode, 149
Mother-of-pearl, 109
Murfreesboro, Arkansas, 216
Muscovite, 39*, 41, 47, 58, 71, 74, 78, 87, 88, 91, 276, 295
Muscovy glass, 91
Muzo, Colombia, 223
Mwadui, Tanzania, 215
Mweza Range (Rhodesia), 223

Naica, Mexico, 14, 109
National Science Foundation, 38
Natrolite, 97, 98, 101, 102, 295
Nepheline, 44*, 50, 295
Nepheline syenite, 50, 65
Nephrite, 252*, 257, 258
Newcomen engine, 219
New South Wales, Australia, 149
Newry, Maine, 78, 240
Nicander, 194
Niccolite, 151, 153, 165*, 168, 295
Nichols, Thomas, 184, 195, 196
Nickel, 38, 142, 151, 152, 153, 157, 158, 159, 160, 168, 193, 194, 195, 245, 269, 270, 277
Nicol, William, 106
Nicol prism, 12, 106
Niobium, 72, 87, 91
Nishapur, Iran, 253
Nizhni-Tagil, Siberia, 157, 172, 260
Norite, 158
Noselite, 50

Obsidian, 29, 48
Obsidian Cliff, 48
Old Faithful geyser, 248
Old Stone Age, 29, 227
"Oldowan culture," 26
Olduvai Gorge, 26, 29, 95
Oligoclase, 47, 48
Olivine, 36, 41, 43*, 45, 46, 48, 49, 58, 83*, 158, 256*, 260, 261*, 262*, 263, 295
Onyx, 229, 245
  Mexican, 104
Opal, 208, 232*, 243*, 244*, 246-248, 295
Operation Mohole, 38
Optical goniometer, 12

Orange Tree State, 150
Oriental alabaster, 104
Orpiment, 23*, 163, 295
Orthoclase, 47, 77, 89*, 295
Orthorhombic crystal system, 15
Osmium, 157
Ovalle, Chile, 250
Oxygen, 41

Pacific Coast Borax Company, 138
Painted Desert, 204
Pala, California, 85
Paleolithic culture, 29
Palermo Mine (New Hampshire), 79
Palisades, 95
Palissy, Bernard, 35
Palladium, 157
Pallasites, 262*, 263
Parahilgardite, 124
Paris, Maine, 82
Pearl, 12, 109, 208
Pearl spar, 104
Pecos County, Texas, 38
Pectolite, 99, 101*, 186, 295
Pegmatites, 71-91, 236, 260, 269
Peridot, 83*, 256*, 260, 263
Peridotites, 49, 58, 158
Peristerite, 77
Perm, Russia, 122
Petalite, 79, 80, 91
Peter's Point, Nova Scotia, 100
Petroleum, 12, 38, 291
Phenacite, 88
Phenocrysts, 43*, 45
Phillipsite, 95, 97
Phlogopite, 39*, 57, 73, 78, 79, 88, 295
Phosphate, 65, 79
Phosphorescence, 185-186
Phosphorus, 41, 65, 79
Physical properties, 19-25
Piezoelectricity, 232
Pike's Peak, Colorado, 77, 88
Pitchblende, 152, 153, 269, 270, 271
Pizarro, Francisco, 146
Plagioclase feldspar, 47-48, 50*, 74, 77
Plasma, 245
Plaster of Paris, 128

Platinum, 102, 142, 143, 157-158, 160, 194, 270, 295
Plato, 194
Playas, 132
Pliny, 11, 14, 194, 195, 197, 208, 229, 235, 246, 287
Pollucite, 79, 80, 91
Porcelain, 28*, 35
Porphyritic texture, 45
"Porphyry copper ores," 178
Port Radium, 270, 271
Portland, Connecticut, 73
Potash, 129-130
Potash feldspar, 47, 48, 74, 77, 87
Potassium, 41, 45, 46, 95, 130, 160
Potassium chloride, 118, 121
Potsherd, 33
Potter's Ridge, 35
Pottery, 33-36
Pough, F.H., 86
Prase, 245
Precious stones, 208-225
Precious topaz, 86
Prehnite, 98, 99, 156, 295
Premier mine, 214-215
Priceite, 132
Prinsloo, Joachim, 214
Proustite, 167, 168
Psilomelane, 254
Pudding stone, 51
Pueblo Bonita, 253
Pumice, 48
Put-in-Bay (Lake Erie), 117
Pyrargyrite, 167, 168, 295
Pyrite, 22, 24*, 60, 124, 146, 165*, 171, 193, 198, 199*, 201, 206-207, 223, 249, 251*, 256*, 277, 289*, 295
Pyroclastic rocks, 45
Pyrolusite, 254, 295
Pyromorphite, 175, 295
Pyrope, 258, 259, 295
Pyroxene, 41, 45, 46, 48, 49, 58, 158, 257
Pyroxenite, 49, 158
Pyrrhotite, 57, 158, 165*, 168, 193, 201-202, 207, 295

Quartz, 11, 14, 15, 20, 21, 22, 29, 31, 36, 41, 42, 47, 48, 50, 51, 52, 54, 57, 58, 62*, 66, 71, 73, 74, 75*, 76*, 77-78, 85,

301

86, 87, 94*, 99, 115, 124, 131, 144*, 146, 149, 165*, 170*, 186, 196, 198, 199*, 204, 207, 223, 226*, 227–248, 256*, 258, 275, 276, 277, 295
   coarse-crystalline, 236
   fine-grained, 240
   rose, 78, 240
   rutilated, 63, 240
   smoky, 239–240
Quartzite, 52, 57, 59, 64*, 158
Quebec, Canada, 97, 284
Queensland, Australia, 247
Queretaro, Mexico, 247
Quicksilver, 144

Radioactive dating, 29
Radioactivity, 264, 267–269
Radium, 152, 153, 264, 269, 271
Realgar, 161*, 163, 295
Reduction, 202
Refractive index, 20, 21
Rhenium, 159
Rhodes, Cecil, 213
Rhodium, 157
Rhodochrosite, 254, 256*, 295
Rhodolite, 259
Rhodonite, 19, 253–254, 255*, 295
Rhombohedron, 10*
Rhyolite, 48
Rio Tinto district (Spain), 207
Ritter Hot Springs, 92, 95
Rock crystal, 20, 78, 236
Rock salt, 121, 123
Rocks:
   extrusive, 42, 45, 48
   igneous, 42–50, 57, 60, 71, 72, 158
   intrusive, 42, 45, 47, 48
   metamorphic, 52–59, 66
   pyroclastic, 45
   sedimentary, 50–52, 103–117
   types of, 37
   vesicular, 95
   weathering of, 50–52
Rosiclare, Illinois, 115
Roxbury conglomerate, 51
Rubellite, 82
Rubidium, 101
Ruby, 20, 57, 63, 208, 210*, 217*, 219–222, 249, 259
Rulandus, 197

Ruskin, 82
Rust, 197
Rustenburg, South Africa, 158
Ruthenium, 157
Rutile, 60, 61*, 63–64, 79, 219, 226*, 240, 295
Ryan Borax Works (Death Valley), 135*

Sacramento, California, 143
Sahara Desert, 51
St. Austell, Cornwall, England, 276, 277
St. Gotthard, Switzerland, 65, 66, 196
Salmon River Mountains, 42
Salt, 15 fn., 21, 52, 101, 118, 120*, 121, 122*, 122–126, 131–132, 291
"Salt-glazed stoneware," 35
Salzburg Alps, 223
Samarksite, 80
Sand, 35, 52, 112*, 236, 260
Sandawana, Rhodesia, 223
Sandstones, 38, 51–52, 55*, 57, 58, 59, 117, 198, 236, 271
   deltaic, 52
   flexible, 58
   fluviatile, 52
Sanford Lake, New York, 48, 203
Sapphire, 20, 63, 208, 218*, 219–222
Sarabit Elkhadem (Sinai Peninsula), 250
Sard, 242
Sardinia, 176
Sardion, 228
Saskatchewan, Canada, 130
Sassolite, 132, 134
Satin spar, 129
Saugus, Massachusetts, 205
Saxony, 123, 151, 168
Scalenohedron, 10*
Scheelite, 20, 186, 295
Schist, 54, 57, 66, 98
Schneeberg, Germany, 104, 151
Schroeckingerite, 270
Scintillation counter, 269
Scolecite, 97
Scott, Sir Walter, 246
Sea, minerals from the, 118–130
Searles Lake (California), 130, 140–141

Sea water, composition of, 118
Sedimentary rocks, 50–52
   crystals in, 103–117
Selenite, 109, 111, 129
Serpentine, 58, 158, 205, 258, 284, 295
Serra de Cordoba (Argentina), 71
Shale, 35, 38, 52, 54, 57, 59, 117, 178
Shikoku, Japan, 164
Shinkolobwe, Congo, 270
Sialk, Iran, 154
Siderite, 88, 90*, 190, 193, 198, 206, 277, 295
Silica, 50, 58, 59, 95, 204, 231, 247
   opaline, 241
Silicon, 34, 41, 160, 231, 235
Silliman, Benjamin, 82
Sillimanite, 54, 57, 66, 67, 295
"Sillimanite Group," 66–68
Siltstones, 52
Silver, 32, 36, 88, 99, 101, 111, 142, 143, 150–153, 154, 156, 160, 163, 164, 167–168, 177, 264, 269, 270, 295
Silver chloride, 151
Sinai Peninsula, 250
Sinter, siliceous, 248
Siskiyou County, California, 257
Skutterudite, 151
"Smalt," 35
Smaragdus, 87
Smith, F. M., 138
Smithsonite, 173*, 174*, 176, 295
Snake River, 55*, 92
Soda ash, 141
Sodalite, 44*, 50, 250, 295
Soddyite, 266*
Sodium, 41, 45, 46, 95, 97, 101, 102, 160
Sodium bromide, 118
Sodium chloride, 118, 121
Sodium sulfate, 121
Sodium chloride, see Salt
Solutré, France, 29
South Crofte mine, 280
South Point, Kau, Hawaii, 261*
Specific gravity, 22
Specularite, 197, 204
Sperrylite, 158

Spessartite, 256*, 258, 259, 295
Sphalerite, 10*, 88, 104, 162*, 164, 165*, 175, 176, 177, 181, 201, 277, 295
Sphene, 47, 48, 60, 61*, 64–65, 295
Spinel, 57, 59, 219, 220, 222, 295
Spodumene, 14, 73, 79, 80, 82, 83*, 84*, 85, 91, 295
Sprudelstein, 103
Stalactites, 104, 105*, 176
Stalagmites, 104, 105*
Star Lake (New York), 197
Starlite, 63
Star of India, The, 218*
"Star of South Africa," 215
Stassfurt, Germany, 122, 126, 130
Staurolite, 54, 57, 65*, 66, 66*, 295
Steel, 187
Steno, Nicolaus, 12, 15, 229, 230
Steno's law, 12
Sterling Hill, New Jersey, 182
Stibnite, 161*, 164, 295
Stilbite, 93*, 95, 96*, 97, 98, 295
Stiles, Ed, 138
Stirling, Lord, 182
Stokes, Sir George, 185
Stone Mountain (Georgia), 47
Stoneware, 35
Stony Point, North Carolina, 85
Strickland's quarry (Connecticut), 73
Strontianite, 117, 296
Strontium, 41, 101, 117
Suckow, John, 140
Sudbury, Ontario, 158, 168
"Sugar Loaf" (Rio de Janeiro), 47
Sulfosalts, 163
Sulfur, 41, 124, 163, 207, 283, 286*, 287–288, 289*, 290*, 291, 296
Sunstone, 77
Superior, Lake, 98
Surinam, 178
Surtsey, island of, 100
Sutter, John, 143
Sverdlovsk, Russia, 223, 254
Syenite, 48, 50, 54
Sylvanite, 149, 296
Sylvinite, 130
Sylvite, 52, 127, 129–130, 141

Symmetry, 16, 19
  axis of, 16
  center of, 19
  plane of, 16

Taconite, 205
Takovaya (Ural Mountains), 85
Talc, 21, 54, 296
Talmage, J. E., 111
Tantalite, 91
Tantalum, 72, 79, 87, 91
Taos, New Mexico, 66
Teflon, 115
Tehuantepec, Mexico, 291
Tell Halaf, Iraq, 154
Tennantite, 296
Tenorite, 278
Teplitz, Czechoslovakia, 259
Tetragonal crystal system, 15
Tetrahedrite, 167, 296
Thallium, 101, 164
Theophrastus, 11, 194, 195, 228, 235
Thermoluminescence, 112
Thetford Mines, 284
Thomsonite, 97
Thorium, 64, 72
Tibet, 134
"Tiger's eye," 85, 282*, 284
Tin, 72, 79, 159, 272, 275–280
"Tincal," 134
Tincalconite, 132, 133, 139*
Ting Tang mine, 279
Titanium, 41, 65, 79, 160, 203, 219
Titanium dioxide, 63, 79
Tivoli, Italy, 104
Topaz, 14, 16, 21, 73, 79, 80, 86, 88, 90*, 220, 221, 239, 276, 277, 296
  precious, 86
Topsham, Maine, 86
Torbernite, 79, 270
Tourmaline, 73, 76*, 79, 80, 82, 219, 226*, 276, 277, 296
Tracey, Percival White, 214
Trachyte, 48
Transvaal, South Africa, 68, 149, 223, 284
Trapezohedrons, 98
Traprock, 49, 92–102
Travertine, 103
Tremolite, 59, 257, 284, 296

Triboluminescence, 112
Triclinic crystal system, 15
Triphylite, 79, 80
Tri-State district, 104, 109, 201
Troilite, 195*
Trona, 141
Trona, California, 141
Tsumeb, Southwest Africa, 172, 176
Tuff, 45
Tulase County, California, 257
Tungsten, 72, 160, 186, 270, 276, 277
Turkey-fat ore, 176
Turkestan, 257
Turquoise, 28*, 36, 250, 252*, 253, 253*, 296
Tuscan Springs, California, 137
Tuscany, 134
"Twenty-mule team," 135*, 138
Tyrolite, 108*
Tyuyamunite, 270

Ulexite, 132, 133, 136*, 137, 138, 139*, 140
Ultramarine, 250
"Underclay," 35
Ungava trough (Canada), 205
Unit cells, 12, 14–15, 16, 19
United States Borax Company, 140
Ural Mountains, 14, 73, 77, 85, 86, 88, 157, 172, 197, 223, 284
Uraninite, 152, 153, 265*, 269, 270, 271, 277, 296
Uranium, 79, 152, 153, 186, 264, 266*, 267–271, 272, 276, 278
Uranophane, 270
Uruguay, 239
Uvarovite, 69*, 258, 259, 260, 296

Vanadinite, 174*, 175, 296
Vanadium, 271
Vandenbrandeite, 266*
Variscite, 253, 255*
Värutrask, Sweden, 73, 91
Veatch, John A., 134, 137
Venus' hairstone, 63, 240
Vermiculite, 146
Verneuil, A., 221
Verneuil process, 63, 221–222
Vesicles, 95